ADVANCES IN ENVIRONMENTAL PSYCHOLOGY

Volume 3
Energy: Psychological Perspectives

Edited by **ANDREW BAUM**
JEROME E. SINGER
Uniformed Services University of the Health Sciences

LEA LAWRENCE ERLBAUM ASSOCIATES, PUBLISHERS
1981 Hillsdale, New Jersey

Lawrence Erlbaum Associates, Inc., Publishers
365 Broadway
Hillsdale, New Jersey 07642

Library of Congress Cataloging in Publication Data

Main entry under title:

Energy conservation.

(Advances in environmental psychology; v. 3)
Includes bibliographies and indexes.
1. Energy conservation—Psychological aspects.
I. Baum, Andrew. II. Singer, Jerome E. III. Series.
TJ163.3.E542 333.79'16'019 81-282
ISBN 0-89859-063-9 AACR1

Printed in the United States of America

Contents

Preface

It is not difficult to understand why interest in psychological aspects of energy production, consumption, and conservation has increased during the past decade. Shortages, soaring prices, and the like have served to underscore the finite nature of our fossil fuel supplies, and accidents such as the one at Three Mile Island or any one of a number of coal mine cave-ins serve as a reminder of the dangers of fuel production. The stresses of living or working near power plants, the problems inherent in reducing fuel consumption, and the many applications of behavioral science to these issues have all received attention of late. This volume, the third in a series, is concerned with these issues—the ways in which psychology can contribute to understanding and helping to solve what has generally been termed an "energy crisis."

Most research in this area has been concerned with strategies for improving conservation and reducing consumption of energy. As McClelland and Canter discuss in their chapter, a number of problems are inherent in such an effort. Matters of design, choice of dependent variables, adequate consideration of intervening variables, and so on, are difficult but important questions with which researchers must deal. As they point out, however, this is but the tip of the iceberg; the issues involved in studying conservation promotion are complex and consistently defy easy solution.

Some of these problems are explored in the context of research programs that have been directed towards conservation. Brechner and Linder, for example, present a social trap analysis, based on the tragedy of the commons, the classic confrontation between private and public welfare. That is, a strategy which maximizes each individual's short term gains may ultimately prove disastrous to the collectivity and therefore to each individual's long term best interests. By considering the energy situation in terms of a social trap, useful analyses are provided.

Reichel and Geller, in their chapter, present a different view of research on energy conservation, applying behavioral technology to the problem of modifying consumption behavior. Primarily concerned with automobile use, they describe some effective strategies for encouraging energy-saving transportation. Seligman, Becker, and Darley describe a related strategy in their discussion of feedback in reducing home energy consumption. However, they point out that feedback is not sufficient to "solve" the energy crisis—dissemination of information and the use of more energy-efficient machines must also be taken into account.

Geller, in another chapter, discusses an issue related to energy conservation—recycling, waste reduction, and resource recovery. Again, applied behavioral analyses are shown to be effective in increasing recycling and recovery.

The last two chapters in the volume depart from the themes of conservation and consumption and deal instead with issues surrounding alternative forms of energy production. Slovic, Fischhoff, and Lichtenstein discuss the interplay of public attitudes and energy policy by focusing on the perception of hazard and risk. By reviewing their research, they highlight the necessity of basing our notions of management of alternative energy systems on an understanding of how people gauge the risks involved. Finally, Sundstrom, Lounsbury, DeVault, and Peele discuss the role of attitudes towards nuclear power in public acceptance of the construction of a nuclear facility. Acceptance is clearly related to risk perception and analysis of potential consequences and benefits.

This volume on energy is the third in the Environmental Psychology Advances series. Volume 1 dealt with urban environments, Volume 2 with perceived control. This book also is thematic, focusing on psychological approaches to the analysis and rectification of energy problems. Despite this common thread, the chapters vary in both outlook and approach. They also differ in their emphasis on theoretical analysis and empirical findings. The reasons for this diversity are related to the fact that our psychological knowledge of the energy situation is still in its infancy.

It is obvious that the energy question is really a congeries of problems. If one imagines a huge matrix of many, many dimensions, the enormous complexity of the area can be realized. One dimension of the matrix would be sources of energy; a second dimension could be facets of the problem, e.g. use or conservation; a third dimension could be aspects of dealing with the problem such as attitude change, economic incentives, etc. Even the simplest of structures would possess a very large number of cells. Ideally, a book such as this would in some fashion touch upon most of them and their interactions. This volume is far from complete in that sense. And it is not just limitations of time, space and money that have kept it from expanding. For despite the current importance and interest in energy, there has not been that much activity in the area on the part of psychological researchers. The matrix has been very sparsely investigated. This book represents, then, not so much a summary or integration because, unfortunately, the field is not well enough developed for that. Rather it is a progress report of the most fully developed work to date. It is our hope that it will provide impetus and models for future researchers in the area.

We would like to thank Brenda Thew and Marcia Gillcrist for their help in producing this volume.

ANDREW BAUM
JEROME E. SINGER

1 Psychological Research on Energy Conservation: Context, Approaches, Methods

Lou McClelland
University of Colorado, Boulder
Rachelle J. Canter
Behavioral Research Institute, Boulder

Since the 1973 oil embargo heralded an "energy crisis" for the United States, interest in energy conservation has risen sharply. For psychologists in the subdisciplines of attitudes and attitude change, behavior modification, program evaluation, and group processes, the energy crisis has provided an opportunity for the development of behavioral solutions to an important social problem. In this chapter, a general psychological context for individual energy use is first presented to guide later discussion. Next, four major psychological approaches to promoting energy conservation are described along with studies evaluating each. Finally, methods typically used in energy conservation research, and their attendant methodological problems, are discussed.

CONTEXT

The consumption of energy—electricity, heating fuels, and gasoline—is the by-product of a wide variety of actions by which individuals achieve other, desired goals. Energy use is thus virtually never an end in itself but rather a means toward many ends. This basic aspect of energy use differentiates it from many behaviors investigated by psychologists and is largely responsible for its particular psychological context. This context is displayed in schematic form in Fig. 1.1.

1. *Individual benefits.* Energy use has many immediate positive consequences or benefits, including comfort (e.g., stable interior temperatures), time savings (especially in transportation), convenience (e.g., using one's own car to

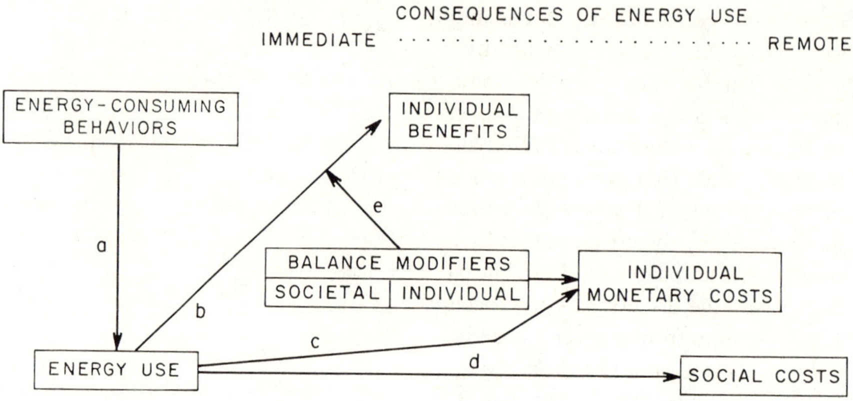

FIG. 1.1. The psychological context of individual energy use.

go places, rather than carpools or mass transit), freedom from environmental restraints (lighting at night), status (from acquisition of manufactured goods), health, and safety. These benefits usually occur simultaneously with, or very soon after, the energy is consumed.

2. *Individual monetary costs.* Energy use has few immediate negative consequences. Acquisition of an energy source, such as an electric hookup, is rarely difficult in the United States, and maintenance of ongoing supplies becomes an issue only in cases of emergency. Of course, energy does cost money. However, most forms of energy are comparatively inexpensive in the United States, and the cost of energy is often small compared to that of substitutes. For example, human power and materials to clean an oven as thoroughly as electricity does would cost several times as much. Furthermore, the monetary costs of energy are not usually paid by the individual immediately but are delayed for several weeks until the utility or credit card bill arrives.

3. *Social costs.* The nonmonetary costs of energy use—air pollution, environmental destruction, political instability, future resource shortages—are even more remote from the individual. They are distant in time, or societal in their effects, or both. Thus, the only negative consequence of energy use readily apparent to most individuals is cost, and even cost is significantly further removed from the time and source of energy use than are the positive consequences.

The pattern of consequences associated with energy use poses a dilemma for energy conservation programs, because behavior is to a large extent controlled by its consequences, with immediate consequences the most effective control. As Bandura (1969) notes: "Findings from experimental studies demonstrate that behavioral changes proceed most effectively when reinforcement is made immediately contingent upon the behavior one wishes to foster [p. 231]."

4. *Balance modifiers.* Some individual conservation actions do not actually save energy but facilitate savings by modifying the balance between energy use and its benefits. For example, home insulation and lightweight, fuel-efficient automobiles allow a reduction in energy use with no loss of benefits; furthermore, the reductions are relatively permanent, requiring no maintenance by the individual. However, such actions do involve outlays of money, time and effort. Actions of this type, hereafter called "balance-modifying actions," can be instituted by society as well as by individuals. For example, if a portion of the through-traffic lanes and desirable parking in an area are reserved for carpools and buses, the drivers of single-occupant automobiles will derive significantly fewer benefits from a given expenditure of energy.

5. *Energy-consuming behaviors.* A second type of conservation action—driving less, lowering the thermostat, taking shorter showers, removing decorative and unneeded lights—reduces both energy use and its benefits. These actions usually require frequent repetition but may sometimes have relatively permanent effects. All such actions require adaptation to reduced positive benefits; however, the benefits will decline proportionately less to the extent that the initial use was wasteful or inefficient. Because individual balance-modifying actions lead to no loss of benefits and are relatively permanent, whereas changes in energy-consuming actions may result in definite loss of benefits and are relatively temporary, different methods of encouraging the two are usually required.

Psychological research on energy conservation is ultimately intended to help in the search for methods of reducing consumption. Individuals now have many opportunities for reducing consumption; more will become available as energy-saving and -producing technologies become more accessible. The behavior changes associated with energy conservation require that the user look for opportunities to conserve, recognize those opportunities when they appear, and take advantage of them when recognized. These opportunities take place within what might be called the "*energy incentive structure*"—the total context of benefits, short-term costs, and remote costs of energy use.

Four major psychological approaches to energy conservation are discussed next; these stem directly from the schema outlined in Fig. 1.1. The first approach explicates the links between components of the energy incentive structure (labeled *a, b, c, d, e*). Announcements, explanations, and demonstrations articulate the relationships between such things as energy use and its long-term environmental and societal costs or between individual actions and energy consumption. Changing basic attitudes toward energy use is the target of the second approach. The third approach relies on actions by societal agents such as governments, large employers, and industry to change the balance between energy use and individual benefits. Finally, the monetary costs to individuals of energy use and/or of balance-modifying actions can be changed through, for example, energy price increases and legislation reducing the costs of energy-saving technologies. A somewhat different psychological framework for energy conservation research, and a review of studies, is presented by Winett and Neale (1979).

All four approaches are grounded in the psychology of attitude and behavior change. Bandura (1969) has categorized attitude and behavior change strategies into three types, to be used independently or in combination. The first, a *belief-oriented strategy,* uses new information or persuasion to change attitudes and, ultimately, behavior. The goal is behavior change within the current incentive structure rather than change of that structure. The first energy conservation approach, explication, typifies the belief-oriented strategy. Altering the evaluation of and behavior toward an attitude object such as energy use is the focus of the second, *affect-oriented strategy*. This strategy relies on techniques of classical conditioning to change attitudes and corresponds to the second energy conservation approach. The third, *behavior-oriented strategy,* is premised on the notion of the functional value of behavior, or the idea that changing behavior requires a change in its consequences. The third and fourth approaches to energy conservation focus on modifying the immediate benefits and monetary cost components of the current incentive structure, respectively, and thus are clearly behavior-oriented strategies.

In the next section each of the four major approaches to energy conservation is discussed in detail. Underlying psychological assumptions, examples of actual use, evaluations, and theoretical and practical problems are considered. Each approach attempts to change the individual's energy-consuming behavior so that less total energy is consumed, or to bring about individual balance-modifying actions so that less energy is needed to achieve the same level of benefits. Although all the approaches are described as they affect individuals, they can also be aimed at households, groups, and even communities. These variations in target do not alter the processes involved and thus are not considered separately.

APPROACHES TO ENCOURAGING ENERGY CONSERVATION

Explicating the Incentive Structure

Consumers of energy often do not accurately perceive the consequences of their energy-consuming actions (Milstein, 1977). A homeowner, for example, may underestimate the proportion of the home's energy use attributable to heating hot water, may not realize that an investment in insulation would pay for itself quite soon, and may think that a constant thermostat setting will use less energy than frequent setbacks. The informational approach presumes that a motive to save energy (or to save money, or to reduce long-term societal costs of energy use) already exists but that the individual's knowledge and understanding of the incentive structure are inaccurate, so that energy-saving opportunities are not recognized or taken. A large number of studies (see Bandura, 1969, for a review) document the significant role of awareness of response-reinforcement contingencies in facilitating behavior change.

The informational approach depends entirely on logic, explanation, demonstration; the current incentive structure is *explicated—not changed*. Examples are common both in research and in conservation programs administered by governments, utilities, and public interest groups. The general approach can be divided into six subapproaches corresponding to different components of the incentive structure. Though often combined, these subapproaches are nevertheless theoretically distinct and are considered separately hereafter. They are listed in Table 1.1. The subapproaches and their particular problems are considered first, followed by a discussion of issues common to all informational approaches.

1. *Link a: How-to-save energy information.* Most of the publicity materials from government agencies, utility companies, and public interest groups do little more than catalog ways in which the individual can modify energy-using behaviors to reduce consumption. Such materials both serve as reminders of commonly known methods (drive at 55 mph, dry clothes outside) and attempt to combat myths many people hold about energy efficiencies (turn out fluorescent lights whenever they are not needed, because the starter does not require more energy than operation; set back the thermostat at night, because reheating in the morning does not consume more energy).

Whereas anecdotal evidence exists that how-to-save information, especially if not known previously, can change behavior, systematic evidence is rare; see Hayes and Cone (1977) and Heberlein (1975) for small-scale studies. Problems

TABLE 1.1
Types of Informational Approaches[a]

1. *Link a*: Explaining how to modify one's energy-using actions so as to consume less energy: Driving at 55 mph uses less gasoline.
2. *Link b*: Persuading that certain benefits can be achieved without as much energy: 70° interior temperatures are not necessary for good health.
3. *Link c*: Explaining how much energy actually costs: Over the next 10 years, electricity to run this freezer will cost $\$X$.
4. *Feedback*: Teaching the individual about links *a, b,* and *c* via more immediate and/or specific feedback than is currently available: Continuous display energy monitors, daily feedback on heating fuel use.
5. *Link d*: Explicating the long-term societal costs of energy use: If the United States runs short of energy, great economic displacements will result.
6. *Link e*: Explaining how balance-modifying actions can change the benefits received from a given amount of energy: With a humidifier, the interior winter temperature can be lowered a few degrees with no change in comfort.

[a]The six subapproaches correspond to components of the incentive structure diagramed in Fig. 1.1.

of broad scope and lack of control over implementation impede evaluations of mass information campaigns by governments and utility companies. In addition, most campaigns and research designs combine how-to-save information with incentives, feedback, or cost information, so that the effects of the how-to-save-energy information cannot be isolated. The work of Craig and McCann (1977a, 1977b, 1978a, 1978b) is an exception. They evaluated how-to-save-energy information sent by mail to several randomly selected samples of Consolidated Edison of New York customers. The information varied on several dimensions suggested by prior research on persuasive communications. Customers receiving information consumed less energy than controls in one experiment; novel messages and high-credibility sources were judged to enhance message effectiveness.

Obstacles to the success of how-to-save-energy campaigns are the same as those facing any informational approach to behavior change. If the individual does not care about energy savings or lower energy costs, how-to-save information will be ignored. Given a receptive audience, the information must be *understandable,* and it must be *believable*. The latter becomes especially critical when myth countering or radical changes are advocated; unfortunately, government, business, and utility companies do not receive especially high believability ratings from energy consumers (Milstein, 1976). To be most useful, the information should be *tailored* to the individual's own situation: type of home, type of heating system, family size, and so forth. Finally, the "list-of-tips" nature of many informational communications leads to a danger of *overload:* So many directives are given that the consumer cannot pick out those that are most important and so adopts none. Any competent attitude researcher faced with a typical how-to-save-energy publication could easily make dozens of suggestions for improvement along the lines already discussed; systematic evaluation of such changes might well be a worthwhile research activity.

2. *Link b: Explanations of the current energy-benefits balance.* This subapproach is based on the idea that intensive energy users have become accustomed to using energy inefficiently and do not realize what nonmonetary benefits could result from reduced energy use. Examples include persuading an individual that at 55 mph, one feels more relaxed and enjoys the scenery more than at 70 mph; or that at 65°, one feels more energetic and healthier than at 75°. The converse is to persuade that heavy energy use has hidden costs of which the individual is not aware, such as automobile servicing, parking, and insurance.

This approach has not been evaluated separately from other informational methods. The critical problems associated with it center on believability: Why is 65° now regarded as healthful whereas no mention was made of this in the past? Believability might be augmented by using appropriate experts as communication sources—for example, physicians on interior temperatures, bacteriologists on cold-water washing.

3. *Link c: Cost information*. This subapproach includes information on the direct costs of energy use: "An electric dryer run on high for an hour costs *X*¢"; and on the monetary savings resulting from reduced use: "A change in thermostat setting from 70° to 60° every night for 8 hours will save *Y*% of the heating bill." It is commonly used by government and utility company publicity programs and is generally accompanied by how-to-save information. No direct evaluations of the subapproach have been conducted to date.

The effectiveness of cost information campaigns may be hampered by problems of accuracy, believability, and motivating power. First, accurate cost information depends on the individual's particular situation—including housing type, climate, and current conservation practices—and on current and future energy prices, which vary widely. This makes production of accurate, easy-to-understand cost information extremely difficult. Perhaps because of the accuracy problem, believability is again an issue. Consumers who fail to see the savings advertised for a given conservation practice on their next energy bill may come to doubt the validity of all cost information. And because of the aggregation of charges over a month and variations in weather, energy prices, and other energy uses, the effect on the bill may not be apparent at all.

A third problem lies in the motivating power of even the most accurate, believable information on energy costs. There is evidence that consumers think most energy-consuming activities cost *more* than they actually do (Carter, 1977; Craig & McCann, 1977b). In this case, successful communication of cost information could *increase* use. Even when this is not the case, how costs should be expressed for maximum motivating power is an issue: Should costs be given per trip, dryer load, hour, week month, year? The greater the aggregation, the greater the amount of money involved, but the more remote the costs will seem.

4. *Feedback*. A general problem limiting the efficacy of information in inducing behavior change is its believability and applicability to the individual's own situation. Feedback programs attempt to convey thc same ideas as information campaigns—a full understanding of how energy-using behaviors affect consumption and/or costs. As such, feedback is a special case of the three subapproaches considered earlier. Feedback differs from these in that the information provided is completely personalized, which should make that information both command more attention and convey the nature of the current incentive structure more effectively.

Purchasers of energy receive some feedback already—the bill. Special feedback programs have moved in two directions to amplify the information usually given on the bill: first, by providing comparisons to past use, expected use, or others' use; second, by increasing the frequency and/or decreasing the delay between energy use and feedback; see Baldwin (1977) for a complete consideration of feedback possibilities. Only the first approach has been tested on a large scale, when utility companies have changed billing formats to include comparisons to past consumption levels. Russo (1977) reports an average systemwide reduction of 2% to 3% for one company adopting this approach. Numerous experimental studies have evaluated the effects of continuous (Becker, Seligman, & Darley, 1979; McClelland & Cook, 1979–80), daily (Becker, 1978; Becker & Seligman, 1978; Bittle, Valesano, & Thaler, 1979, 1979–80; Collins, 1977; Seligman & Darley, 1977; Winett, Neale, & Grier, 1979; Winett, Neale, Williams, Yokley, & Kauder, 1978–79), weekly (Winett, Kagel, Battalio, & Winkler, 1978), or monthly (Hayes & Cone, in press; Seaver & Patterson, 1976) feedback of various forms. The results suggest that both daily and monthly feedback can reduce energy consumption for heating, cooling, lights, and appliances.

Although feedback has been demonstrated as relatively effective in promoting energy conservation, obstacles to its widespread use and issues concerning how best to maximize its effectiveness still remain. First, feedback cannot be obtained without cost and effort, either on the part of the consumer or an outside agent. This limitation makes many of the feedback systems tested impractical for general use. The U.S. Department of Energy is currently testing an in-home monitor with continuous and cumulative daily and monthly displays; the device would retail for under $100 (DOE Office of Conservation, 1978a).

Second, the possible combinations of terms (dollars, kilowatt-hours; per day, month, hour, year), comparisons (to own past use, a norm, others' use, prediction based on weather), frequency, lag, and cumulating period are virtually endless. Both theory and empirical evidence suggest that variations along these parameters influence the effectiveness of feedback; however, it is not yet clear which combinations most effectively reduce which types of energy use.

Third is the question of how confounding influences should be treated. Energy use is the result both of the individual's behavior and of less controllable factors such as weather, equipment efficiency, and traffic density. Energy bills are also a function of energy prices, which may change frequently. The issue of the treatment of confounding influences is dual. First, should such influences be removed from feedback? They should be removed if the intent is the education of the consumer as to the effects of behavior on energy use; they should not be removed if the intent is an accurate depiction of energy use or costs for a given period, or of how the opportunity for energy savings varies with external events such as weather. Second, if removal of confounding influences is deemed desirable, is it practical or even possible? Removal of even so large and regular an effect as that of weather on heating fuel consumption is not as straightforward as it might seem, especially when the period in question is as short as a day. The consumption of control units has often been used in research to remove confounding influences; however, this may not be an acceptable long-term solution. The problem of confounding influences is critical because feedback that does not correspond roughly to what the recipient expects is not believable and may be totally ineffective (Becker & Seligman, 1978).

5. *Link d: Explication of long-term social costs.* Most informational material on conservation makes some reference to the long-term societal impact of energy depletion. Direct tests of the effectiveness of this approach alone have not been made. However, social learning theory suggests that the remoteness of the consequences from the individual will severely limit the effectiveness of the approach. As Stern and Kirkpatrick (1977) observe, "Organisms normally do not forego a large individual reward in order to avoid a small share of a delayed punishment [p. 11]." For example, even very personal long-term consequences such as death from lung cancer are not powerful deterrents for many people, and the consequences of energy depletion are far less personal. Societal consequences—even if immediate, such as air pollution from auto exhaust—can be dismissed via the mechanism of diffusion of responsibility ("Someone else will do it, so I needn't"; see Darley & Latané, 1968) or a "commons dilemma" argument ("My conserving won't have any effect unless everyone else conserves, so I won't bother"; see Hardin, 1968; and Platt, 1973). The issue of believability is also of major import. Information about future and societal consequences is by definition not verifiable by the individual; rather, the "experts" must be trusted. When experts disagree, as has been common of late, the believability of all suffers.

6. *Link e: Explanations of balance-modifying acts.* This subapproach, along with how-to-save information, is the most widely used of the general informational approaches. Information explaining how balance-modifying actions can change the benefits received from a given amount of energy (or the energy required for a particular benefit) usually advocates a purchase. Small automobiles, home insulation, energy-efficient appliances, humidifiers, solar panels, heat pumps, and even down comforters are all sold with the argument that the purchaser can reduce energy use with no loss in comfort. Utility company campaigns for insulation and hot-water savers and the U.S. Department of Energy's "Project Conserve" (computerized information on insulation needs for homeowners; Stroh, 1974) and "Project Payback" (a life-cycle costing and payback-period advertising and product-labeling program; DOE Office of Conservation, 1978b) are other examples of the subapproach.

Advertising campaigns and large-scale governmental programs are extremely difficult to evaluate. A systematic assessment of the effects of Project Payback is planned for 1980; otherwise, we can only assume from the prevalence of advertising geared to energy savings that industry has found this subapproach effective.

The problems facing this subapproach have all been discussed earlier; they center primarily on believability (especially when the information comes from a vested interest), the accuracy of cost information, and the most effective terms for contrasting future savings to current costs (Cunningham & Lopreato, 1977; Hanna, 1978; Montgomery & Leonard-Barton, 1977). Some programs—Project Conserve, home energy audits performed by utilities—attempt to alleviate the believability and accuracy problems by tailoring information to the individual's own situation. This, of course, requires more effort on the part of both the agency and the consumer, who must provide or allow collection of detailed information on housing status, current energy use, and so forth.

Summary. Three major obstacles to success face the six subapproaches to explicating the incentive structure. These obstacles are not unique to energy conservation programs, but are common to all attempts to modify attitudes and behaviors with information. They are: (1) the believability and acceptability of recommended actions, (2) dependence on existing motives, and (3) commanding the attention and involvement of the audience so that attitude change is translated into behavior change.

Decades of research in social psychology have demonstrated that communicator credibility is critical to attitude change and that any message too discrepant from the audience's initial view will be rejected (see McGuire, 1969, for a review). A message from the president to set thermostats at 63°, for example, might be rejected because the source is not expert (Will 63° really require less energy than 65°? Is it healthy?), not respected, or has a vested interest; or because 63° is so far from current settings as to be unthinkable.

A second major obstacle is that all the subapproaches depend on existing motives for saving money, conserving energy, and/or preventing the negative long-term societal costs of energy use. If these motives are weak, or if the balance of positive and negative consequences in the current incentive structure is insufficient to arouse them, explication cannot change behavior no matter how well it is done. Evidence from surveys (Cunningham & Lopreato, 1977; Frankena, 1977; Milstein, 1976, 1977) suggests that saving energy per se and preventing long-term societal consequences of energy depletion are not salient motives for U.S. consumers. Furthermore, the same surveys also indicate that even money—in the amounts currently available as a result of conservation—is not a universally strong motivator, especially for high-income groups.

The third obstacle centers on the importance of the individual's thinking about his or her current energy conservation practices. Behavior change via information is a multistage process: The individual must attend to the message, believe and accept it, and act upon it. The lack of strong existing motives is only one reason why action may not follow from belief and acceptance. A second reason is that the individual is unable to act because of lack of money, lack of control over environmental conditions, and the like. A third is that the individual considers him- or herself already in compliance with the message by doing all possible to conserve. This attitude may also prevent the person from attending to the message at all.

If the final stage of the behavior change process—the step between belief-acceptability and action—is not complete, an apparent "attitude-behavior inconsistency" results; such inconsistencies are notoriously common in energy research. For example, Milstein (1977) found that in 1976, U.S. homeowners reported acceptance of 66° as a daytime thermostat setting but that actual settings averaged 70°. Initial attention to the persuasive message and translation of attitude change into action should both be facilitated by information that is clearly relevant to the individual's own situation, such as that provided by Project Conserve, utility company home energy audits, and feedback programs.

Changing Attitudes Toward Energy Use

As already noted, the success of explicating the energy incentive structure in promoting conservation depends on already existing motives to save energy or money. Whereas instilling motivation to save money is certainly beyond the reach of any conservation effort, instilling a motive to save energy per se may not be. The slogans "Don't be fuelish" and "Save a kilowatt," appeals for conservation by respected national figures, and characterizations of conservation as a patriotic duty are examples. Such messages are most often accompanied by additional information on how to save energy and money and have not been evaluated directly.

Changing attitudes is not guaranteed to change behavior but might at least facilitate the process if positive intrinsic consequences were to become associated with conservation. However, changing attitudes without reference to their consequences is extremely difficult, requiring perhaps hundreds of repetitions of the message in what is essentially a classical conditioning paradigm. Furthermore, the "good" (person, message) associated with conservation must be unequivocally good to at least a majority of the target audience, and achieving this may be difficult, also. In sum, the "affect-oriented" strategy for changing attitudes toward energy use in and of itself can play only a small role in effective conservation programs.

Changing the Energy-Benefits Balance

The last two approaches to promoting energy conservation assume that the current incentive structure supports waste and must be *changed*—not simply explicated. These approaches—changing the energy-benefits balance and changing the monetary costs of energy and associated actions—are quite similar formally. In the first, nonmonetary consequences of energy consumption—both positive and negative—are modified, whereas the second considers monetary costs exclusively.

Changes in the energy-benefits balance can only be made by agents with the ability to control some part of the energy-benefits relationship: national and local government, large employers, the energy technology industry. Examples of this approach include a city's providing more comfortable buses and special carpool lanes to make energy-efficient transport more comfortable and faster; a national or state program of rationing or licensing for new or increased energy use to make the costs in time and effort of such use greater; an employer's relegating all single-occupant automobile parking to a distant, inconvenient lot to decrease the benefits of that mode of transport; and industry's marketing of new products that require less energy or effort to achieve a given benefit, such as heat pumps and clock thermostats.

Because energy researchers usually do not qualify as "agents controlling part of the energy-benefits relationship," this approach is difficult to evaluate systematically. People's self-reports on how they would act if the incentive structure were to change are notoriously unreliable; the only valid evaluations of the approach, therefore, are those of actual large-scale implementations. Unfortunately, these are most often evaluated by such measures as bus ridership, citizen complaints, or sales, rather than by total energy use (Seligman, 1978).

The obstacles to success for the approach of modifying the energy-benefits balance depend in part on whether the actions taken primarily increase the benefits of energy-efficient behaviors or reduce the benefits of energy waste.

When a government, employer, or industry acts so as to increase the available benefits of energy-efficient behaviors—for example, by improving the comfort and frequency of buses—the action can have no effect on the behavior of individuals unless it is known to them. Those already engaged in the efficient behavior, bus riders, will receive increased benefits and therefore will be more likely to maintain the behavior. Communicating the new incentive structure to potential participants, single-occupant automobile drivers, entails all the problems of attention, believability, and acceptance discussed under informational approaches. Sometimes special benefits such as entertainment and refreshments on buses and/or vastly reduced costs (free rides) are instituted temporarily to attract new users, who can then "see for themselves" the increased benefits. For example, after an 11-month program of no-fare bus service ended, the Regional Transportation District of metropolitan Denver lost only 20% of the new riders gained during the program.

The obstacles to success are quite different when an action taken by an employer or government reduces the benefits of widespread energy-wasting behaviors. Examples include an employer's moving noncarpool parking far from the building and restriction of single-occupant vehicles to fewer lanes in rush hour. The change in incentive structure will certainly draw attention, although attempts to arouse interest before the change actually occurs often fail. Instead, the problem is one of reactance (Brehm, 1966), whereby people respond negatively to any compulsory change made without their consent no matter what the change might be. A Federal Energy Administration survey (Milstein, 1977) found an overwhelming preference for voluntary rather than compulsory conservation measures, and for incentives for saving rather than penalties for waste.

Although prior consultation with and approval from those to be affected by a compulsory change—via public hearings, notices, persuasive campaigns—is the obvious antidote to negative response, drawing those likely to respond most vehemently into the debate-and-approval process is often difficult. A second factor important in determining how well reduced benefits for energy use will be received is equity (Milstein, 1977; Montgomery & Leonard-Barton, 1977). If the "sacrifices" required by an energy conservation measure are seen as affecting various population segments equally, the measures will be more readily accepted.

A final obstacle to the success of modifications of the energy-benefits balance is strictly practical. Any such changes involve costs—in money, effort, negative public reactions, and possible noncompliance—for the agent of change. These costs are often substantial and may be uncertain as well; for example, political reactions to rationing, licensing, and the like are difficult to predict. In addition, the ultimate benefit to the agent from reduced energy use may also be uncertain (will carpool lanes really encourage more carpooling?), and the potential benefits are usually societal in scope and distant in time, at least beyond the next election. Given these conditions, the reluctance of agencies, especially governmental agencies, to implement modifications in the energy-benefits balance is understandable.

Changing the Monetary Costs of Energy Use and Associated Actions

Like the previous approach, this one has two opposite sides—increasing the price of energy so as to reward conservation and penalize waste, and decreasing the real price of balance-modifying actions through tax incentives, subsidies, loan policies, and the like. This approach assumes that the current incentive structure is understood but is inadequate to encourage conservation.

Examples of this approach abound. Utility companies can raise rates, reverse declining block rate structures, institute "lifeline" rates to benefit small consumers, finance purchase and installation of energy-saving technologies. Governmental authorities can rebate the costs of energy-saving technologies, tax energy-using devices according to their efficiencies, subsidize the manufacturers of conservation technologies, tax energy use per se (as in the gasoline tax), loan money for conservation purchases, and fine excessive energy use. Financial institutions can set loan rates for homes and vehicles according to their energy efficiency and finance the purchase of conservation technologies. Finally, employers, landlords, and heads of institutions can pass on charges for energy use, fine for waste, and give rebates and awards for conservation.

Evidence. With the exception of the last category (actions by landlords, institution heads, and employers), the actions just listed are so broad in scope as to preclude well-controlled evaluation. Three types of nonexperimental evidence are relevant to the assessment of the effectiveness of the monetary costs approach.

First, price-elasticity studies by economists estimate quantitatively the effects of changes in the price of energy. Such studies show a small but consistent negative effect of price increases on consumption (Lehman & Warren, 1978; Montgomery & Leonard-Barton, 1977; Schneider, 1978).

Second, self-reported reactions to price changes and tax incentives indicate that many consumers anticipate taking progressively more conservation actions as prices and incentives increase (Cunningham & Lopreato, 1977; Honnold & Nelson, 1978; Milstein, 1977). Third, comparisons can sometimes be made between cities or service areas with different monetary costs for energy use or waste.

Some experimental evidence on the effects of positive monetary incentives on conservation is available (Foxx & Hake, 1977; Hayes & Cone, 1977; Kohlenberg, Phillips, & Proctor, 1976; McClelland & Belsten, 1979–80; McClelland & Cook, 1980b; Newsom & Makranczy, 1977–78; Slavin & Wodarski, 1977; Slavin, Wodarski, & Blackburn, in press; Walker, 1979; Winett, Kagel, Battalio, & Winkler, 1978; Winett, Kaiser, & Haberkorn, 1976–77; Winett & Nietzel, 1975). This evidence primarily concerns the effects of rebates and prizes on energy use by small numbers of apartment residents, dormitory residents, and homeowners. The studies indicate that positive financial incentives can lead to some conservation, at least for a limited time (3 to 10 weeks). However, the monies distributed have usually exceeded the value of the energy saved; the effects have often faded over time; and many residents seem unaware of or uninterested in the monies available. Two examples from research conducted by the authors illustrate the potentials and pitfalls of reliance on positive financial incentives to encourage conservation. McClelland and Cook (1980b) conducted a conservation contest among residents of four groups of master-metered apartments in University of Colorado married-student housing. The group using the least natural gas (relative to its own baseline) in each of six 2-week contests won $80 to use however it determined. Although consumption was significantly below normal for the entire contest, savings diminished from 10% in the first to 4% in the last 2-week period. Residents reported adopting some conserving behaviors but had virtually no knowledge of contest results despite extensive publicity. Practical problems encountered included the size of the prize monies ($780) relative to the savings produced ($488) and difficulties in accurately predicting "normal" use, even with complex regression analyses.

McClelland and Belsten (1979–80) aided residents of a 1000-student dormitory at the University of Colorado in conducting a "conservation challenge" program. In each of three 2-week periods, the residents could win, for the entire dormitory's use, $85, $215, or $335 for approximately 5%, 10%, or 15% reductions in electricity use from baseline. Consumption dropped by over 10%, saving $350 in electricity costs; $385 in prizes were awarded. Although a randomly selected group of residents surveyed later knew about the program, almost none knew even approximately how much money had been available. The incentive seemed to motivate the actions of dormitory staff and student government rather than to affect resident behavior directly.

Combinations of positive monetary incentives for conservation and negative incentives for waste should, according to theory, be even more effective than positive incentives alone. Natural experiments testing this notion occur in rental multifamily housing when fixed "utilities-included" rents are replaced by direct resident payment of energy costs; or, in other words, when a property is converted from master to individual metering. Such conversions lead to energy savings averaging 22% for energy used for cooling, lights, and appliances, 5% for energy used for heating and hot water (McClelland, 1980).

Because such metering conversions are not always practical or economic, the authors have investigated for the U.S. Department of Energy nontechnological methods of encouraging conservation in master-metered housing. The most promising method to date is one in which the resident pays a monthly utility charge representing the resident's share of consumption for the entire property. Shares are apportioned according to square feet occupied. Residents thereby receive both benefit for conservation and penalty for waste. This system leads to savings averaging 5% to 8% for all energy functions. (See McClelland, 1980, for complete results.) The savings produced by conversions to individual metering and by pro rata allocation of group costs exceed those from rebate and prize systems in both size and permanence, thereby supporting the superiority of positive–negative incentive combinations.

Problems. The obstacles to success facing the monetary-costs approach mirror those discussed in the preceding section. First, negative public reactions, reactance, noncompliance to penalty or fine systems, even boycotts may follow price increases, especially if the effects of the price changes are not seen as equitable. Because utilities and energy suppliers generally benefit from price increases, this is a serious problem. Second, positive tax incentives and price decreases in conservation technologies may not be noticed or understood by many consumers, as was the case in the two studies already described. Third, even price increases may not be noticed if they are gradual in nature and hidden in bills determined by weather and consumption as well as by price. Such gradual hidden increases will be more acceptable but undoubtedly less effective than sharp increases in encouraging conservation.

Fourth, changes in the monetary consequences of energy use may not be sufficient to reduce energy consumption. Although money is commonly cited as "the most effective incentive in inducing a consumer to conserve energy" (Milstein, 1976, p. 9), the nonmonetary benefits of energy use—comfort, convenience, status—may be strong enough to make moderate price increases ineffective in prompting conservation (Milstein, 1977; Seligman, Kriss, Darley, Fazio,

Becker, & Pryor, 1979). And even when money is the primary motivator, the consumer must know what actions to take to conserve before changes in monetary costs can be effective.

The practical obstacles to implementation of the monetary-costs approach are as formidable as those to the approach of modifying the energy-benefits balance. When prices are increased, especially by taxes, strong negative reactions are inevitable. Positive tax incentives and subsidies cost the acting agency money and may also lead to negative public reactions because they thereby raise other costs or are inequitable. The acting agent thus must be relatively certain that the goal of the action is worthwhile; that the action will lead to the goal; that the costs in effort, money, and public reaction will not exceed the benefit; and that no one population segment will benefit or suffer unduly.

METHODS

The ultimate purpose of psychological research on energy conservation is the identification of effective means of promoting conservation. Such research thus typically tests one or more approaches to energy conservation, with comparison of resulting energy use to before-treatment levels, to use under an alternate approach, or to a control with no program. Field experiments and quasi-experimental program evaluations are the research methods of choice (Committee on Measurement of Energy Consumption, 1977; Seligman, 1978; Winett, 1976). In this section, four commonly used research methods are described, examples of each are given, and the central methodological problems of each are considered. Following that, common issues concerning measurement of energy use, problems of generalization, and limitations on researcher control are discussed.

Four Common Research Strategies

With virtually all energy conservation research performed in the field, methodological differences across studies center on means of enlisting participants and of assigning participants to conditions. The bulk of empirical research to date can be classified into four strategies, each with its own methodological problems: (1) true experiments with random assignment of randomly selected participants to experimental and control conditions; (2) true experiments with random assignment of volunteer participants; (3) quasi experiments in which controls are recruited for self-selected or arbitrarily chosen experimental participants; (4) designs with self, statistical, or no controls.

True Experiments. In this method, participants randomly drawn from a specified population such as an apartment building, utility company service area, or neighborhood or from the entire population are randomly assigned to a control condition and one or more experimental conditions. Participants are not informed that they are in an "experiment" or, usually, that their energy use is to be monitored. For example, McClelland and Cook (1979–80) evaluated the effects of energy monitors providing residents of 25 North Carolina homes with continuous feedback on electricity use. The homes were arbitrarily selected to receive monitors (as original equipment) by the builder from a subdivision of 101 homes. Data were collected from utility company archives after 18 months of occupancy. Households with monitors consumed 12% less electricity over an 11-month period than those without. Other studies using true experimental designs are Heberlein (1975), Seaver and Patterson (1976), and Slavin and Wodarski (1977). Researchers using this strategy face problems concerning invasion of privacy, informed consent, and confidentiality.

True Experiments with Volunteers. When experimental participants must play an active role in the conservation program rather than passively receiving information or rebates, they must give their consent to participate. Many studies, especially of feedback procedures, therefore use volunteer participants. For example, Becker and Seligman (1978) recruited 40 participants from identical three-bedroom townhouses in a New Jersey housing development. All volunteers agreed to monitoring of electricity use and to random assignment to one of four experimental conditions or to a control condition. Recruitment methods range from newspaper ads (Winett & Nietzel, 1975) to door-to-door campaigns designed to attract a large proportion of the population (Winett, Neale, & Grier, 1979, recruited 60% of the households approached). Other studies using true experimental designs with volunteer participants include Becker (1978), Becker and Seligman (1978), Seligman and Darley (1977), Winett, Kagel, Battalio, and Winkler (1978), and Winett, Neale, Williams, Yokley, and Kauder (1978–79).

Methodological problems of this strategy stem directly from the use of volunteers. First, would the experimental treatments affect those who did not volunteer in the same way as they affected study participants? Will the treatments still be effective when they are not considered experimental? Nietzel and Winett (1977) present survey results indicating that volunteer participants may differ radically from other community residents on a number of dimensions concerning conservation. Second, are control participants who know that their energy use is being monitored by a conservation experiment truly adequate? They may well provide adequate control for influences on energy use such as weather, as is suggested by Winett, Neale, and Grier's (1979) finding of no differences in use by their volunteer controls and a nonvolunteer comparison group. However, whether voluntary controls would respond to such influences as national conservation appeals or price increases in the same way as nonvolunteers has yet to be established.

Quasi Experiments. In this strategy, experimental participants may volunteer or may be selected for convenience in implementing the treatments or in obtaining energy-use data. Control participants are recruited explicitly as controls or, more commonly, do not know of their role as controls. (For a complete discussion of many "quasi-experimental" designs and their hazards, see Campbell & Stanley, 1963.) For example, Hayes and Cone (1977) compared energy use in four volunteer apartments receiving incentives, information, and feedback on energy use for the entire 80-unit building. Other examples of this method include Foxx and Hake (1977), McClelland and Belsten (1979–80), McClelland and Cook (1980a, 1980b), Newsom and Makranczy (1977–78), Slavin, Wodarski, and Blackburn (in press), Walker (1979), Winett (1977–78), and Winett, Kaiser, and Haberkorn (1976–77). Russo (1977) compared systemwide energy use for a utility company with feedback on monthly bills with use for companies without feedback. The planned evaluation of the Department of Energy's Project Payback (DOE Office of Conservation, 1978b) can also be considered a quasi experiment. Market statistics from cities with special life-cycle cost-advertising programs are compared to figures from similar cities without such programs.

The quasi-experimental method incorporates the strengths and weaknesses of true experiments and of use of volunteers. The question of how the treatments would have affected nonvolunteers remains open when experimental participants are self-selected. Issues of consent and invasion of privacy surround the use of unknowing control participants. When the "participants" for whom energy use is measured are apartment buildings, cities, or utility company service areas, however, the ethical issues diminish in magnitude and are far outweighed by the clarity a control–experimental comparison brings to the interpretation of results.

Designs with Self, Statistical, or No Controls. Baseline–treatment–return-to-baseline designs commonly used in behavior modification studies use self-controls to guard against the permanent effects of external influences coincident with the experimental treatment. Examples of such ABA designs in conservation research include Kohlenberg, Phillips, and Proctor (1976) and Palmer, Lloyd, and Lloyd (1977). Such designs cannot guard against temporary influences on energy use such as weather or vacations, nor can such designs usually be employed in large-scale evaluations of government or employer programs. In these cases, statistical controls—modeling temporary effects during a baseline period, then removing them statistically—may be used. For example, McClelland and Belsten (1979–80) used one semester of weekly electricity data to model the effect of weather on dormitory consumption, then used the resulting regression equation to remove weather effects from data from the following three semesters. Lehman and Warren (1978) used a similar method to estimate the price elasticity of natural gas. The obvious interpretational question facing such studies is the influence of temporary or permanent effects that are unknown and therefore not removed.

Issues Common to All Strategies

No matter what methodological strategy is employed, the psychologist in energy conservation research faces three common problems. These concern measurement of the dependent variable, the generalization of results to new settings, and practical limitations on empirical investigations. Each is considered in turn.

Measurement of the Dependent Variable. Although large-scale program evaluations sometimes use "input" variables such as citizen participation as success criteria, the primary dependent variable in conservation research must be energy use (Seligman, 1978). Because energy is consumed by multi-individual units and because consumption is simply a means to other desired goals, generation of meaningful data on energy use in the laboratory is impossible. Self-reports of current energy use or of use in hypothetical situations are equally suspect because of the aura of social desirability surrounding energy conservation and because most people are ignorant of their own consumption (Milstein, 1977). The researcher must therefore collect energy-use data in the field or depend on archival records collected for other purposes. Neither method is without problems.

Collecting consumption data in the field can assure the researcher a high degree of validity but may be quite expensive in both time (personpower to read meters at fixed intervals) and equipment costs (meters or recording devices if not already present). Most critically, long periods of frequent monitoring are usually required to yield enough data for stable estimates of consumption levels.

The use of archival records collected by utility companies or by the consuming units themselves circumvents many of the practical problems involved in researcher-collected data. However, the unit of aggregation—city, utility company service area, building—may be larger than is desired, and the frequency of observations much lower. Utility company records are often time-consuming to obtain and sometimes costly as well; they frequently contain substantial amounts of missing data and may not include certain information critical for analysis—exact dates of the meter readings, for example. The accuracy of all records must be carefully checked by the researcher: In one series of 80 monthly, utility company readings for an apartment building, the authors discovered 12 consumption figures in error by 20% or more. Because one meter-reading error causes compensating errors in the data for adjacent periods—one month is higher than expected, the next too low—such errors can be discovered with careful use of regression analyses.

Generalization of the Results. The goal of research on energy conservation is the identification of methods of encouraging conservation that will be effective in settings other than those in which they were tested. This goal can be complicated by the fact that many diverse types of behaviors can lead to reduced energy use. If a treatment succeeds by prompting behavior A and that behavior is not available to occupants of a new setting, possibly because it has already occurred there, the treatment may not succeed in the new setting. Thus, generalization of results will be aided by knowledge of the mechanisms by which various approaches or treatments operate.

For example, a household feedback program might lead to reduced energy use via lower thermostat settings, installation of insulation, or the occupants' spending less time at home. Self-reports from participants and/or direct observations of behavior could help pinpoint the mechanism for conservation in this case. The researcher must be most cautious about using behavioral measures alone, without direct measures of energy use, because unobserved compensating changes can occur: Participants might drive less but faster, have the lights off because they are working elsewhere, keep the thermostat lower but use the oven to heat the kitchen.

Practical Limitations. The researcher engaged in true or quasi-field experiments meets practical constraints both with respect to the energy conservation approaches that can be tested and the availability of participants. Of the four approaches to energy conservation described in the first portion of this chapter, only the first two—information and changing attitudes toward energy use—can be implemented fully by a researcher acting alone. The other two approaches—changing the energy-benefits balance and changing monetary costs—require instigation by or at least full cooperation of more powerful societal agents, especially when the effects on any party are negative. Furthermore, temporary implementation of these approaches is often not feasible because of anticipated negative public reactions to their introduction or withdrawal.

Participant availability is limited by the fact that energy conservation is seldom controlled directly by individuals, but rather by larger units such as households, business firms, buildings, and other groups. The larger the units of study, of course, the fewer the units in the population. Also, determining the mechanisms by which a successful program has operated can be very difficult when each consuming unit contains many individuals. A second limitation on participant availability becomes important when random assignment is not possible: Participants in different conditions should be matched on opportunity to reduce energy use. Such matching has been attempted by McClelland and Belsten (1979–80), McClelland and Cook (1980a), Newsom and Makranczy (1977–78), Walker (1978), and Winett and Nietzel (1975). It is important because participants with high levels of energy waste can reduce consumption by a given proportion more easily than participants with low waste levels. This phenomenon and its disruptive effect on the researcher's ability to interpret differential energy reductions has been noted by Becker (1978), McClelland and Cook (1980a), Winett, Kagel, Battalio, and Winkler (1978), and Winett and Nietzel (1975). For example, in a rebate information study, Winett et al. (1978) contrasted the 10 households with the largest energy reductions to the 10 with the largest increases. A major difference between the groups was that many of the reducers took common conservation actions after the experimental procedures began, whereas many of the increasers had done so before the study began. The necessity for matching on opportunity to reduce energy use is a practical limitation on participant availability because estimating opportunity is very difficult and because appropriate matches may simply not exist within the population.

CONCLUSIONS

Energy conservation research presents a myriad of theoretical, practical, and methodological challenges to its practitioners. In the theoretical arena, the researcher must carefully assess the current incentive structure or psychological context surrounding the behavior to be encouraged or modified. Does the behavior lead directly to reduced energy use, or does it change the balance between energy use and its benefits? What are its immediate and long-term, individual and societal, positive and negative consequences? Are consumers knowledgeable about these consequences? If not, simple explication of the incentive structure might lead to conservation, provided that the persuasive information is believable, acceptable, and commanding of attention. Conversely, if consumers are aware of the current incentive structure but nevertheless still waste energy, what changes in that structure could be made to encourage conservation? The researcher's theoretical knowledge of attitude and behavior change techniques, and of the psychological context of energy use, must be drawn upon in planning potentially effective strategies for promoting particular types of energy conservation.

The psychologist in energy conservation research must also consider several practical and methodological issues. Can the strategy of choice be implemented by the researcher acting alone? If not, what existing agencies can be enlisted to help? Can evaluations of the planned or existing conservation programs of governments and utility companies yield useful information on the value of the strategies used? How can such evaluations best be performed? Problems of measuring energy consumption, recruiting participants, locating appropriate controls, and determining the mechanisms of change so that results can be generalized to new settings will challenge the researcher's methodological skills.

Although the development and evaluation of successful energy conservation strategies pose many challenges to the psychologist, one should not conclude that such research is not worthwhile, for several reasons. First, the necessity for energy conservation in the United States is clear at present and will only increase in the future. Second, behavioral strategies are potentially cost-effective means of promoting energy conservation. Even small reductions in consumption can be very significant; when aggregated over thousands of households, businesses, or communities, seemingly small reductions produce large energy savings. Third, behavioral approaches will always play a role in conservation programs because no technological approach can be of value unless and until people adopt it. Finally, the true and quasi-field experiments required to test conservation strategies should provide a balance to more traditional laboratory research on attitude and behavior change, thereby shedding new light on the psychological processes involved. In sum, research on energy conservation is a socially and scientifically worthwhile activity for psychologists.

REFERENCES

Baldwin, F. D. Meters, bills and the bathroom scale. *Public Utilities Fortnightly*, 1977, *99*, 11–17.

Bandura, A. *Principles of behavior modification*. New York: Holt, Rinehart & Winston, 1969.

Becker, L. J. Joint effect of feedback and goal setting on performance: A field study of residential energy conservation. *Journal of Applied Psychology*, 1978, *63*, 428–433.

Becker, L. J., & Seligman, C. Reducing air conditioning waste by signaling it is cool outside. *Personality and Social Psychology Bulletin*, 1978, *4*, 412–415.

Becker, L. J., Seligman, C., & Darley, J. M. *Psychological strategies to reduce energy consumption: Project summary report* (PU/CEES 90). The Center for Energy and Environmental Studies, Princeton University, Princeton, NJ 08544, June 1979.

Bittle, R. G., Valesano, R., & Thaler, G. The effects of daily cost feedback on residential electricity consumption. *Behavior Modification*, 1979, *3*, 187–202.

Bittle, R. G., Valesano, R., & Thaler, G. The effects of daily feedback on residential electricity as a function of usage level and type of feedback information. *Journal of Environmental Systems*, 1979–80, *9*, 275–287.

Brehm, J. S. *A theory of psychological reactance*. New York: Academic Press, 1966.

Campbell, D., & Stanley, J. *Experimental and quasi-experimental designs for research*. Chicago: Rand McNally, 1963.

Carter, J. L. Saving energy, saving dollars: How young couples—North and South—do it. *Redbook*, November 1977, pp. 118; 121; 184–186; 188.

Collins, L. D. Social comparison and objective standard feedback as means of reducing residential energy consumption. *Dissertation Abstracts International*, 1977, *37*, (11-B), 5878.

Committee on Measurement of Energy Consumption, Assembly of Behavioral and Social Sciences, National Research Council. *Energy consumption measurement: Data needs for public policy* (Section on Field Experimentation). Washington, D.C.: National Academy of Sciences, 1977, 78–85.

Craig, C. S., & McCann, J. M. Communicating energy conservation information to consumers: A field experiment. In B. A. Greenberg & D. A. Bellenger (Eds.), *Contemporary marketing thought*. Chicago: American Marketing Association, 1977. (a)

Craig, C. S., & McCann, J. M. *Marketing energy conservation to residential consumers of electricity: Report to the Federal Energy Administration* (Graduate School of Business and Public Administration paper). Ithaca: Cornell University, 1977. (b)

Craig, C. S., & McCann, J. M. Assessing communication effects on energy conservation. *Journal of Consumer Research,* 1978, *5,* 82–88. (a)

Craig, C. S., & McCann, J. M. The impact of persuasive communications on energy conservation. *Energy Systems and Policy,* 1978, *2,* 433–447. (b)

Cunningham, W. H., & Lopreato, S. C. *Energy use and conservation incentives: A study of the southwestern United States*. New York: Praeger, 1977.

Darley, J., & Latané, B. Bystander intervention in emergencies: Diffusion of responsibility. *Journal of Personality and Social Psychology,* 1968, *8,* 366–383.

Department of Energy, Office of Conservation, personal communication from Lynn Collins, 1978. (a)

Department of Energy, Office of Conservation, personal communication from Bruce Hutton, 1978. (b)

Foxx, R. M., & Hake, D. F. Gasoline conservation: A procedure for measuring and reducing the driving of college students. *Journal of Applied Behavior Analysis,* 1977, *10,* 61–74.

Frankena, F. *Survey research on energy: An annotated bibliography* (Exchange Bibliography No. 1305). Council of Planning Librarians, July 1977.

Hanna, S. Evaluation of energy saving investments. *Journal of Consumer Affairs,* 1978, *12,* 63–75.

Hardin, G. The tragedy of the commons. *Science,* 1968, *162,* 1243–1248.

Hayes, S. C., & Cone, J. D. Reducing residential electricity energy use: Payments, information, and feedback. *Journal of Applied Behavior Analysis,* 1977, *10,* 425–435.

Hayes, S. C., & Cone, J. D. Reduction of residential consumption of electricity through simple monthly feedback. *Journal of Applied Behavioral Analysis,* in press.

Heberlein, T. A. Conservation information: The energy crisis and electricity consumption in an apartment complex. *Energy Systems and Policy,* 1975, *1,* 105–117.

Honnold, J. A., & Nelson, L. D. Public opinion regarding energy conservation. *Journal of Environmental Education,* 1978, *9,* 20–29.

Kohlenberg, R. J., Phillips, T., & Proctor, W. A behavioral analysis of peaking in residential electricity energy consumers. *Journal of Applied Behavior Analysis,* 1976, *9,* 13–18.

Lehman, R. L., & Warren, H. E. Residential natural gas consumption: Evidence that conservation efforts to date have failed. *Science,* 1978, *199,* 879–882.

McClelland, L. *Encouraging energy conservation in multifamily housing: RUBS and other methods of allocating energy costs to residents*. Report to the U.S. Department of Energy. Institute of Behavioral Science, University of Colorado, Boulder, Co. 80309, 1980.

McClelland, L., & Belsten, L. Promoting energy conservation in university dormitories by physical, policy, and resident behavior changes. *Journal of Environmental Systems,* 1979–80, *9,* 29–38.

McClelland, L., & Cook, S. W. Energy conservation effects of continuous in-home feedback in all-electric homes. *Journal of Environmental Systems,* 1979–80, *9,* 169–173.

McClelland, L., & Cook, S. W. Energy conservation in university buildings: Encouraging and evaluating reductions in occupants' electricity use. *Evaluation Review,* 1980, *4,* 119–133. (a)

McClelland, L., & Cook, S. W. Promoting energy conservation in master-metered apartments through financial incentives. *Journal of Applied Social Psychology,* 1980, *10,* 19–31. (b)

McGuire, W. J. The nature of attitudes and attitude change. In G. Lindzey & E. Aronson (Eds.), *The handbook of social psychology* (2nd ed., Vol. 3). Reading, Mass.: Addison-Wesley, 1969.

Milstein, J. S. *Attitudes, knowledge and behavior of American consumers regarding energy conservation with some implications for governmental action*. Washington, D.C.: Federal Energy Administration, October 1976.

Milstein, J. S. *How consumers feel about energy: Attitudes and behavior during the winter and spring of 1976-77*. Washington, D.C.: Federal Energy Administration, June 1977.

Montgomery, D. B., & Leonard-Barton, D. *Toward strategies for marketing home energy conservation* (Graduate School of Business, Research Paper No. 372). Palo Alto, Calif.: Stanford University, 1977.

Newsom, T. J., & Makranczy, U. J. Reducing electricity consumption of residents living in mass-metered dormitory complexes. *Journal of Environmental Systems,* 1977-78, *7,* 215-235.

Nietzel, M. T., & Winett, R. A. Demographics, attitudes, and behavioral responses to important environmental events. *American Journal of Community Psychology,* 1977, *5,* 195-206.

Palmer, M. H., Lloyd, M. E., & Lloyd, K. E. An experimental analysis of electricity conservation procedures. *Journal of Applied Behavior Analysis,* 1977, *10,* 665-672.

Platt, J. Social traps. *American Psychologist,* 1973, *28,* 641-651.

Russo, J. E. A proposal to increase energy conservation through provision of consumption and cost information to consumers. In B. A. Greenberg & D. N. Bellenger (Eds.), *Contemporary marketing thought: 1977 educators' proceedings*. Chicago: American Marketing Association, 1977, 437-442.

Schneider, A. M. A new tax on gasoline: Estimating its effect on consumption. *Science,* 1978, *202,* 755-757.

Seaver, W. B., & Patterson, A. H. Decreasing fuel oil consumption through feedback and social commendation. *Journal of Applied Behavior Analysis,* 1976, *9,* 147-152.

Seligman, C. *Evaluating energy conservation programs*. Paper presented at the meeting of the American Psychological Association, Toronto, 1978.

Seligman, C., & Darley, J. M. Feedback as a means of decreasing residential energy consumption. *Journal of Applied Psychology,* 1977, *62,* 363-368.

Seligman, C., Kriss, M., Darley, J. M., Fazio, R. H., Becker, L. J., & Pryor, J. B. Predicting summer energy consumption from homeowners' attitudes. *Journal of Applied Social Psychology,* 1979, *9,* 70-90.

Slavin, R. E., & Wodarski, J. S. *Using group contingencies to reduce natural gas consumption in master-metered apartments* (Center for Social Organization of Schools, Report No. 232). Baltimore: Johns Hopkins University, 1977.

Slavin, R. E., Wodarski, J. S., & Blackburn, B. L. A group contingency for electricity conservation in master-metered apartments. *Journal of Applied Behavior Analysis,* in press.

Stern, P. C., & Kirkpatrick, E. M. Energy behavior. *Environment,* 1977, *19,* 10-15.

Stroh, R. C. *Project conserve: A pilot project in homeowner energy conservation*. Washington, D.C.: Federal Energy Administration, October 1974. (NTIS No. PB-240 407)

Walker, J. M. Energy demand behavior in a master-metered apartment complex: An experimental analysis. *Journal of Applied Psychology,* 1979, *64,* 190-196.

Winett, R. A. Micro-economic approaches to energy conservation. In J. Kagel & R. C. Battalio, *Experimental methods in economics*. Atlanta: Southeastern Economics Association, 1976.

Winett, R. A. Prompting turning-out lights in unoccupied rooms. *Journal of Environmental Systems,* 1977-78, *7,* 237-241.

Winett, R. A., Kagel, J. H., Battalio, R. C., & Winkler, R. C. Effects of monetary rebates, feedback, and information on residential electricity conservation. *Journal of Applied Psychology,* 1978, *63,* 73-80.

Winett, R. A., Kaiser, S., & Haberkorn, G. The effects of monetary rebates and daily feedback on electricity conservation. *Journal of Environmental Systems,* 1976-77, *6,* 327-339.

Winett, R. A., & Neale, M. S. Psychological framework for energy conservation in buildings: Strategies, outcomes, directions. *Energy and Buildings,* 1979, *2,* 101-116.

Winett, R. A., Neale, M. S., & Grier, H. C. The effects of self-monitoring and feedback on residential electricity consumption. *Journal of Applied Behavior Analysis,* 1979, *12,* 173-184.

Winett, R. A., Neale, M. S., Williams, K., Yokley, J., & Kauder, H. The effects of feedback on residential electricity consumption: Three replications. *Journal of Environmental Systems, 1978-79, 8,* 217-233.

Winett, R. A., & Nietzel, M. T. Behavioral ecology: Contingency management of consumer energy use. *American Journal of Community Psychology,* 1975, *3,* 123-133.

2 A Social Trap Analysis of Energy Distribution Systems

Kevin C. Brechner
California State University, Los Angeles

Darwyn E. Linder
Arizona State University

Along a highway in South Dakota, the Campbell County Soil and Water Conservation District has placed a series of signs. They read:

> Too many cattle
> on too little grass
>
> Like too many cars
> and too little gas

This little quatrain points to the structural similarity of a wide variety of sitations in which a common property resource is exploited for individual benefit until that resource is totally depleted. Many situations share this common structure: vanishing or extinct species of animals, the 1929 stock market crash, overcrowding of citizens band radio frequencies, energy blackouts, overcrowding of our national parks, air pollution resulting from the use of automobiles, the financial difficulties of prepaid health maintenance programs, and the rapid escalation of both energy consumption and the cost of energy. John Platt (1973) labeled these situations *social traps*.

The most well known of these situations is the "tragedy of the commons" (Hardin, 1968). The tragedy is so well known to environmentalists that it may be described here very briefly. In England and in colonial New England, the commons was pasturage for public use rather than private ownership. Each herdsman who used the commons to increase the size of his herd was motivated by the prospect of economic gain. As long as the population of humans and cattle was kept relatively small by disease, famine, and warfare, the system worked quite well. However, humans eventually mastered the environment, and these limitations on growth became ineffective. As the herd size reached the carrying capacity of the commons, the addition of more animals by any herdsman was accompanied by a decrease in the overall productivity of the commons. Eventually, the pasture was destroyed.

In this chapter we describe how the tragedy of the commons as well as the situations already mentioned may be considered social traps. We then describe the development of laboratory analogues of social traps and report some of the results that have been obtained. Finally, we apply the concept of the social trap to patterns of energy flow and resource consumption, and we attempt to suggest ways in which the social trap analogue may be used as a research tool to explore strategies for reducing energy consumption or for promoting energy conservation.

SOCIAL TRAPS AND THE COMMONS DILEMMA

The tragedy of the commons is a social trap (Platt, 1973). According to Platt (1973), social traps refer "to situations in society that contain traps formally like a fishtrap where men or organizations or whole societies get themselves started in some direction or some set of relationships that later prove to be unpleasant, or lethal and that they see no easy way to back out of or to avoid [p. 641]." Platt defined social traps in behavioral or Skinnerian terms. A social trap occurs when an individual performs some behavior in order to obtain an immediate, short-term, positive outcome or reinforcer while the collective actions of all the individuals in the situation lead to a negative outcome in the long term. Thus in the tragedy of the commons, the short-term outcome is the economic advantage to each individual herdsman of a larger herd. The collectively shared long-term negative outcome is the destruction of the commons.

A simple mathematical model of the commons has been provided by Crowe (1969). The utility of an action can be described on a scale that runs from -1 to $+1$, in which the greatest benefit or positive utility is expressed as $+1$, and the greatest cost or negative utility is expressed as -1. The utility of putting an additional animal on the commons even after the carrying capacity has been exceeded is very nearly $+1$ for the individual herdsman, because he will receive the entire benefit from the additional animal. The decrease in productivity of the commons will be shared by all those who use it. Thus only a small fraction of that negative utility will accrue to the herdsman who adds to his herd. At every point in time, even as the destruction of the commons approaches, it is the best interest of each individual to add additional animals. The tragedy of the commons occurs when individuals follow this rational course of action until the commons is overgrazed and destroyed. Making the simple substitution of the word *consequences* for the word *utility* in Crowe's preceding analysis illustrates how the tragedy of the commons is a social trap.

The social trap concept nicely captures the structure of the tragedy of the commons, or the commons dilemma, as it has been called by Dawes (1975). The social trap is, however, a broader concept than the commons dilemma in that social trap analysis can be applied to many situations in which the concept of commons is difficult to apply. All commons dilemmas are social traps, but many social traps cannot be reduced to commons dilemmas. For example, wage–price inflation, in which the demand for higher wages leads to increased prices, which in turn lead to increased demands for higher wages, cannot easily be considered a commons dilemma, but it is clearly a social trap. Some aspects of energy consumption are also social traps but not necessarily commons dilemmas. The problem of peaking in electrical power use is an example. Utility companies must have sufficient generating power to meet peak use demands, which tend to occur between 3:00 P.M. and 8:00 P.M. As peak use expands, the power companies must build additional generating facilities, which increases the cost of electricity. Yet each person is motivated by the availability of electrical appliances and by a concern for individual convenience to use electrical power at a time when it is most needed and most convenient. The result is that the cost of energy continues to increase partly as a function of the need for increased generating capacity that arises only during peak demand hours.

Many situations that we have used as examples of social traps can be analyzed from the perspective of other disciplines. There are, of course, economic analyses of wage–price inflation, the collapse of the financial system, and a number of other phenomena. There are biological analyses of overpopulation and of the limitations of population in some animal species. Although these analyses capture the situations in conceptual terms, they do not suggest solutions in which individual behavior may be changed to avoid the long-term negative consequence. A social trap analysis, however, because it employs psychological and behavioral concepts, may lead more directly to the development of solutions to social traps.

SOCIAL TRAPS AND SUPERIMPOSED SCHEDULES OF REINFORCEMENT

Platt suggested how social traps can be analyzed using an operant framework. Social traps are situations where the short-term consequences for a behavior are positive and the long-term consequences are negative. In the case of social traps, the long-term negative consequence is shared by all the individuals in the group. Platt also described one-person traps, such as smoking cigarettes or overeating, where the long-term consequences of the behavior affect only the individual.

Basic to Platt's analysis was the idea that multiple consequences followed responding. Brechner (1977), in a study that is described later, showed how an analogue of social traps could be constructed in the laboratory by building a reinforcement structure in which each of the consequences was defined by a simple schedule of reinforcement. Fixed-ratio schedules, where a fixed number of responses had to be made, paid off in an immediate reward. Differential reinforcement schedules, where fast responding depleted the number of available reinforcers, defined the long-term negative consequence. These simple schedules were superimposed on each other, so that they all were contingent on a single behavior. Thus, the reinforcement structure was called *superimposed schedules of reinforcement*. Interesting situations other than social traps also can be built with superimposed schedules, particularly when superimposed schedules are combined with concurrent schedules (Brechner, 1978).

With the exception of some animal studies, little work has been done with superimposed schedules. Kelleher's (1966) and Marr's (1969) work with second-order schedules and Weiss and Laties' (1958, 1959, 1963) work with titration schedules to study animals' tolerance to shock can be considered to be superimposed schedules. The studies of self-control by Rachlin (1974) and Ainslie (1975) are also related. Of particular relevance to the study of social traps is Solnick's (1978) paper applying the Ainslie–Rachlin model to humans. In several experiments subjects were given a choice between a small, immediate reward and a large, delayed reward. In both cases, the reward was turning off a very loud noise. Also varied was the length of time (T) that passed between the choice response and the beginning of the consequences. When the time delay was 0, subjects tended to choose the smaller, immediate reward. When T was increased to 7.5 and 15 seconds, subjects more frequently chose the larger, delayed reward. In one study, Solnick found that the effect was mediated by the type of instructional set given to subjects. As described later, instructional sets can play an important role in behavior in a social trap, and they illustrate how cognitive variables can interact with the reinforcement structure to produce behavior.

Superimposed schedules can also be used as an analytic tool to try to approximate the reinforcing contingencies that operate in actual social traps. For example, the increasing cost of electricity for home consumption can be seen as a superimposed schedule of reinforcement. Each day, various responses are emitted that are reinforced by comfort, food, clean clothes, and so on. In operant terms, this is a schedule of continuous reinforcement in which responses to control room temperature, prepare food, and wash clothes are always reinforced. Superimposed on this is a variable-interval schedule of aversive consequences in which the cost of electricity rises through the necessity to meet peak-load power demands. Another example of a social trap based on a superimposed schedule is the problem of illegal aliens streaming into the United States. The aliens themselves are not on a superimposed schedule. They are responding to a quite simple contingency—higher salaries and standards of living in the United States. Those who employ illegal aliens, however, are subject to a superimposed schedule. They receive continuous reinforcement in the form of decreased costs whenever they hire an illegal alien at a lower wage than a citizen could demand. But they also are on a variable-interval schedule of aversive consequences in the form of increased taxes to support social services and government intervention directed at illegal aliens. In both these examples, the schedule of positive reinforcement applies to the individual, whereas the schedule of aversive consequences applies to the entire community. Thus, social trap situations can be analyzed as superimposed schedules of reinforcement. Using superimposed schedules in this manner can be very informative as long as the user realizes that he or she is analyzing the behavior of a complex living human, not a rat or a pigeon or a computer.

ESCAPING FROM SOCIAL TRAPS

One possible solution to social traps, according to Platt, is to change the delay between the behavior and the long-term negative consequences. This is most easily done by making the long-term negative outcome more salient symbolically. This can be done with feedback about current levels of consumption and predicted time until the resource is exhausted. Slogans such as "Give a hoot, don't pollute" and "Every litter bit hurts" attempt to make more salient the consequences of thoughtless disposal of waste materials.

A second solution suggested by Platt is to increase the short-term costs—that is, to make the behavior so costly that it produces a loss instead of a gain. Tickets for exceeding highway speed limits are one example of this kind of additional cost. Exceeding the speed limit may have an attractive short-term benefit for the individual, but in the long run it leads to increased fuel consumption and to more serious injuries and deaths. Effective enforcement of the speed limit adds cost to the behavior of driving as fast as one might wish. Unfortunately, this solution of adding additional costs to the behavior is difficult to implement because it requires surveillance and effective enforcement, which often are not possible. An illustration of this problem can be seen in a modern-day tragedy of the commons occurring in the Southwest United States. The situation involves land jointly owned by the Navajo and Hopi nations. The Hopis charged that Navajos had overgrazed the joint land and, having destroyed that, were invading Hopi lands. "On Low Mountain mesa the overgrazing was so drastic that the hungry herds had been reduced to eating sagebrush. Vast expanses of sagebrush—an almost indestructible plant—had been killed by overgrazing and cedar trees had been cropped as high as sheep and goats could reach" (Thomas, 1972, p. 23). The Hopis formed a grazing committee that imposed daily fines on invading cattle, sheep, and goats. They captured thousands of head of livestock, but the Navajos reportedly crossed fence lines at night when enforcement was difficult. Federal legislation has been drafted to settle the dispute.

Another solution suggested by Platt is to add reinforcers for a competing behavior. Incentives that reward carpooling rather than individual driving are an example. Another example is offering reduced rates for power comsumption at times other than during the peak load. Long-distance telephone rates are structured in this way to encourage calls at times when the lines are less fully used.

A solution identified by Platt, which has been used many times, is to set up a superordinate authority to regulate behavior or the use of the common resource. The Federal Communications Commission, state fish and game agencies, the Federal Reserve Bank, the Federal Aviation Authority, and the Interstate Commerce Commission are examples of superordinate authority to regulate the use of a resource. Curiously, the same newspaper page that described the Hopi–Navajo dispute had a story that related how a regulatory agency can sometimes overregulate. A 65-year-old Navajo man in New Mexico, who ran a tiny, one-man coal mine, was threatened by federal authorities with $60,000 in fines for not installing safety equipment. One requirement was the installation of a telephone between the 50-foot-deep mine and a 5-by-7-foot tool shack near the mine entrance. Navajo Tribal Councilman Harry Tome said: "I don't know what they expect since he's the only guy working the mine. Is he supposed to call the shack from the mine and then run back there and answer the phone?" ("Bureaucracy buries 1-man coal mine," 1972).

Platt also suggested that a solution to the social trap is to change the nature of the long-term consequence. This is almost always a technological solution to the problem. If air pollution can be avoided by the installation of pollution control devices on automobiles, then one of the long-term consequences of automobile use can be avoided. If a cheap, clean, relatively inexhaustible supply of energy can be developed, then the consequences of wasteful fossil fuel consumption can be avoided. However, as Crowe (1969) has pointed out, the tendency of social scientists has been to refrain from doing research on problems that can be identified as social traps, because they expect a technological solution will be found by physical scientists. Physical scientists, on the other hand, see many of our problems that can be analyzed as social traps as requiring political or social solutions rather than technological ones. The end result is that these problems remain unsolved (Crowe, 1969). The social trap concept suggests many kinds of behavioral solutions to these problems as alternatives to technological solutions and should therefore spur research designed to find ways of avoiding the long-term negative consequences, and thus of escaping the trap.

The solutions to social traps suggested by Platt are based primarily on his behavioral analysis. But all solutions except for technological ones are focused on reducing the frequency or the intensity of the response for immediate or short-term benefit. Such changes in behavior may be accomplished by other than the behavioral methods suggested by Platt. For example, superordinate authorities may create a social structure and use norms and social sanctions to decrease the frequency or intensity of the target behavior. The cognitive and motivational factors that maintain the short-term response can be considered. Concepts from social psychology, personality research, and cognitive psychology can be employed to develop strategies to change behavior.

SOCIAL TRAP ANALOGUES

Developing solutions to social traps is the primary purpose of applying a social trap analysis to any situation. But implementing and evaluating solutions to social traps are exceedingly difficult. In most instances the researcher must obtain the cooperation, if not the active participation, of the local, state, or federal unit of government, some regulatory agency, or some large societal institution. The solution that has been developed may then be implemented and a before–after evaluation attempted. However, it is difficult to develop a systematic body of knowledge about the effectiveness of solutions in social traps, because each such instance will be a one-shot case study. Laboratory analogues that have the structure of a social trap or the tragedy of the commons have been developed to allow the systematic and cumulative exploration of solutions to social traps. The first social trap analogue was developed by Brechner (1974/1975, 1977). An electromechanically programmed game was designed to create the tragedy of the commons in the laboratory. Subjects could harvest points from a common pool and use those points to obtain extra experimental credit. Subjects were highly motivated to complete the experimental participation option for their introductory psychology course at Arizona State University. Points that could be converted to extra experimental credit were therefore a highly valued commodity, just as additional cattle are highly valuable to the herdsman, or as electrical power is highly valued for operating air conditioners on hot summer afternoons.

The experimental situation is depicted in Fig. 2.1. Each subject could see the number of points remaining in the pool at all times as indicated by the light that was illuminated in a column of lights. Points could be harvested by depressing the response button mounted on a grey metal box in front of each subject. The box also contained a counter on which each subject's personal total of points was displayed. Thus subjects were always aware of the number of points remaining in the pool and of their own point totals, but were unaware of the number of points that had been obtained by the other subjects in the group.

In an attempt to model biological systems, points were replenished to the pool as a positive function of pool size. When the pool was kept above 75% of its capacity, points were replenished quite rapidly. Points were replenished at a slower rate when the pool was between 50% and 75% of its capacity, more slowly between 25% and 50% of capacity, and very slowly when the pool was less than 25% of its capacity. If the pool was ever depleted to zero, no more points were replenished, and the experimental session ended.

The optimal strategy was to leave the level of the pool high and respond for points as they were replenished. This slow responding maximized the number of points that could be taken during a session. All aspects of the game were explained to subjects except for the variable replenishment rates and the exact details of the conversion table by which points could be exchanged for experimental credit. Subjects were able to observe the differential replenishment rate,

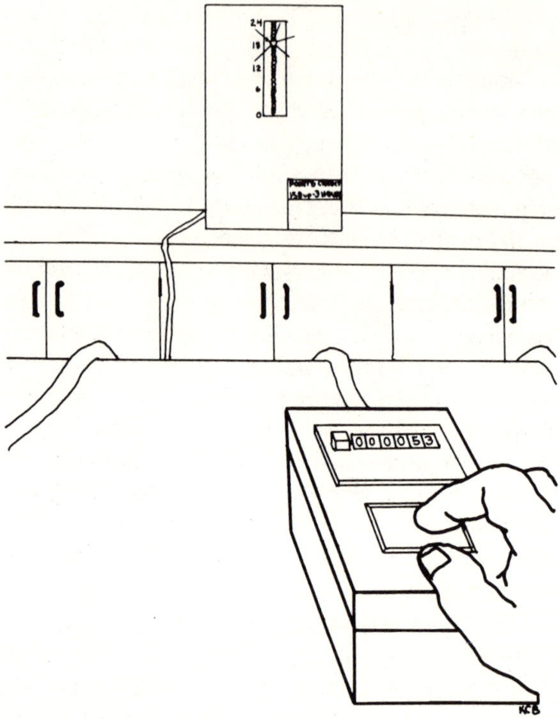

FIG. 2.1. Schematic of the first social trap analog. Three subjects each had their own response unit. Pressing the response unit button 10 times took one point from the common pool and added it to the subject's own counter. The pool display in the background was automatically replenished with points at rates that decreased as the number of points in the pool decreased.

however, because when they entered the experimental room, the pool was in the process of filling, having first been lowered to near zero by the experimenter. After giving the instructions, the experimenter reduced the pool to near zero and told subjects to begin responding when the pool level reached the maximum. Subjects, therefore, had two opportunities to observe the variable replenishment rate. The details of the conversion schedule of points to experimental credit were not revealed to subjects, so that specific subgoals could not be set. Rather, subjects were motivated to obtain as much extra experimental credit as possible.

The group version of the game (Brechner, 1977) was designed to be played by three persons. An individual version (Brechner, 1974/1975) of the game was developed to simulate one-person traps. The individual version was created simply by dividing the replenishment rate function by 3, so that an individual playing the game would be confronted by a per-subject replenishment rate equivalent to that encountered by subjects playing the game in groups. The total capacity of the pool was varied. One-third of the subjects saw a display with a total of 12 lights arranged in a vertical column; one-third saw 24 lights; and one-third saw 48 lights. The replenishment rates were identical for each condition. The optimal strategy was identical for each condition. The rationale for varying pool size was that the small pool size might be more likely to alert subjects to the possibility of pool depletion and the necessity for developing effective management strategies. In other words, the small pool size might make the long-term negative consequence more salient to the subjects. The results indicated that pool size did not have a significant effect on performance. After two-thirds of the subjects could play the game effectively and were able to manage the resource. Additional analysis indicated that the majority of subjects who did deplete the pool were in the smallest-pool condition. Most of the subjects in the 24- and 48-light conditions were able to find an effective strategy.

The group version of this social trap analogue explored two independent variables. The first was communication. Half the groups were allowed free verbal communication with one another. In the other half of the groups, communication was forbidden. The effects of communication on levels of cooperation have been mixed. In some instances communication has facilitated cooperation (Wrightsman, O'Conner, & Baker, 1972), but in other instances communication has had little if any effects on levels of cooperation (Deutsch & Krauss, 1960, 1962). In a social trap analogue, however, communication could be useful in developing norms for appropriate behavior, administering sanctions, or coordinating response rates. The second independent variable was again pool size (24 or 48 lights). If a small total pool capacity does make more salient the possibility of pool depletion, then communication would be especially useful to subjects in the small-pool condition. Thus, it was predicted that the most effective management of the resource would occur when the total pool capacity was small and communication was allowed.

Brechner's social trap analogue allowed the experimenter to collect three closely related but conceptually separable dependent variables. The first was total points harvested by the group. In this experiment, total points harvested could be confounded with total pool capacity. Subjects in the large-pool condition would have an advantage of 24 more points available if they simply depleted the pool as quickly as possible. A second dependent variable was the length of time subjects maintained the pool prior to total depletion. This measure, however, could show equivalent results for a group that maintained the pool at a very low level as well as for a group that maintained the pool at a very high level for the entire experimental session. The third dependent variable was the number of points replenished to the pool until depletion or until the end of the 15-minute experimental session. This measure reflected the effectiveness with which the pool was managed. If the pool was maintained at a high level, then many points were replaced during the experimental session. If the pool was maintained at a very low level, fewer points were replenished. Furthermore, subjects in the large-pool condition did not have an artifactual advantage on this dependent measure. All three dependent variables are of interest, and they are highly intercorrelated in most experiments. Each measure is useful for some specific purposes. Number of points harvested is a useful index of the degree of success experienced by the group. Time until resource depletion reflects the group's ability to avoid total resource depletion. Number of points replenished to the pool is a sensitive index of the effectiveness of resource management.

The results of Brechner's experiment are displayed in Table 2.1. Although an analysis of variance revealed significant main effects for both communication and pool size on the number of points replenished, it was apparent from subsequent analyses that only the small-pool, no-communication condition differed from the other three and showed a consistent pattern of very rapid depletion of the resource. Thus, the initial hypothesis of an interaction between pool size and communication such that the most effective management would occur within a small pool with communication was disconfirmed. The results are, nevertheless, of great interest. First, they indicate that the laboratory analogue can produce behavior similar to that observed in the natural environment. The tragedy of the commons was recreated, especially in the small-pool, no-communication groups. It is important to remember that subjects in these groups were confronted by the same parameters of pool size and replenishment rate that confronted the subjects in the individual version of the analogue. Yet whcreas subjects in those conditions in the individual version were able to manage the resource effectively and obtain many points, subjects in the three-person groups depleted the pool, often in less than one minute. Furthermore, the results of the three-person version of the game indicated that in terms of points replenished, points harvested, and time until depletion, the outcome of the game could be affected by independent-variable manipulations. Thus, the first social trap analogue demonstrated that the tragedy of the commons could be recreated in the laboratory and that independent-variable manipulations could affect whether the tragedy occurred or was averted.

TABLE 2.1
Mean Points Replenished as a Function of Pool Size and Communication[a]

		Pool Size		
		24	*48*	
Communication	Yes	212.84	236.33	*N* cell = 6 groups
	No	40.50	180.00	

[a]From Brechner, K. C. An experimental analysis of social traps. *Journal of Experimental Social Psychology*, 1977, *13*, p. 559. Reproduced by permission of the publisher and author.

The social trap analogue created by Brechner demonstrated the viability of laboratory experiments within the social trap paradigm. However, the electromechanical equipment that Brechner used to create the analogue caused some rather severe limitations on the flexibility of the game, so that certain kinds of experimental variables could not be manipulated. The aged equipment itself was subject to breakdown. One of the limitations was the number of persons who could participate within the experimental setting. The analogue could not be expanded to accommodate more than three people without a complete reconstruction and rewiring. Changing the pool size required about two hours of rewiring. Another problem was that the pool could not be subdivided to allow comparisons between a common property resource and individually held resources. Subdividing the pool would be equivalent to converting an apartment house that is master-metered to a system of metering individual apartments for energy consumption. In addition to the lack of flexibility to accommodate some important independent-variable manipulations, the equipment itself was bulky and difficult to transport to different research locations.

Because of these limitations on the analogue developed by Brechner, several additional analogues have been developed. All of these analogues share the same basic structural characteristics of social traps. Individuals respond to acquire resource units for their own personal gain from a common property resource that is replenished at a rate that is a positive function of the level of the resource. Brief descriptions of these analogues and some of the research results obtained using them are described next.

A social trap analogue in which the resource pool could be subdivided for individual ownership was developed by Cass (1975/1976). Cass used Arizona State University computing-center CRT terminals as display and response devices for subjects. The analogue was programmed in APL for a Univac 1110 computer. Cass manipulated whether or not subjects had feedback about pool level and whether or not the resource was individually owned or collectively shared. The resource consisted of 12 pools of points labeled with letters of the alphabet. In the common-property conditions, subjects could harvest points from any of the 12 pools. In the individual-ownership conditions of Cass's study, subjects could harvest only from pools assigned specifically to them. Attempts to harvest from pools owned by others resulted in a message printed on the CRT screen: "No poaching allowed, try another." Feedback was manipulated either by printing on the screen the number of points remaining in each pool at any point in time or by not showing that information. Thus, in a 2 × 2 factorial design, subjects were either given feedback or feedback was withheld, and they either harvested from a collectively shared resource or from individually owned resource pools.

Once again the behavior of subjects in a social trap analogue mirrored the behavior observed in the tragedy of the commons and other social traps. When the resource was collectively shared, subjects without feedback about the pool level depleted the pool very rapidly. Subjects with feedback about the pool level also depleted it but somewhat more slowly. When the resource was subdivided for individual ownership, feedback had the opposite effect. Without feedback, subjects harvested very conservatively. With feedback, they harvested at an optimal rate—the rate at which the resource was replenished. The social trap analogue developed by Cass once again demonstrated the viability of experimentation within this paradigm. Independent-variable manipulations had strong effects on behavior within the social trap situation. The implication of Cass's study for energy consumption is discussed later.

The social trap analogue developed by Cass was somewhat more flexible than Brechner's analogue but was still limited in that it required a large-scale computer installation for its operation. A more portable, manually operated social trap analogue was developed by Shippee (1978) using a task devised by Edney (1979). In this analogue, resource units were metal hexagonal nuts that subjects manually removed one at a time from a bowl. The bowl rested on a postal scale, so that the experimenter always had a readout of the pool level. Resource units were replenished by the experimenter every 15 seconds. The number replaced was a positive function of the number of units remaining in the pool at the time of replenishment. In the control condition of his experiment, Shippee simply allowed subjects to harvest from the pool and demonstrated once again that the pool was very quickly depleted. In the other conditions, Shippee provided his three-person groups with the option of taking time out from harvesting. This option was an attempt to operationalize the concept of moratorium on harvesting natural resources and resembles the concept of seasons on game animals and moratoriums on the harvesting of species placed on the endangered species list. Although the option in this experiment was for a complete cessation of responding, other options certainly could be used. Options to reduce the intensity or frequency of responding by group agreement might parallel some energy conservation techniques, such as lowering thermostat settings in individual dwellings during the winter and raising them during the summer to reduce consumption of energy for heating and air-conditioning.

The main portion of Shippee's experiment was a 3 × 3 factorial design. One independent variable, type of leader, consisted of three levels—no leader, an appointed leader, and an elected leader. The other independent variable manipulated the manner in which the decision to take time out from responding could be made. Time out could be called unilaterally (but only by the leader in the appointed-leader and elected leader conditions), by agreement of two members of the group, or only by unanimous agreement of all three members. The results of Shippee's experiment indicated that the pool was managed most effectively by groups with an elected leader where the decision to take time out had to be unanimous. This pattern of results is provocative, but replication and elaboration of Shippee's experiment are required before any generalizations to public policy may be attempted. The experiment demonstrated, however, that manipulations other than those emerging from the behavioral perspective adopted by Platt (1973) can have an effect on behavior in a social trap situation. In this instance, the social structure of the group and the decision-making rules were manipulated and had a significant impact on behavior in a social trap analogue.

Shippee's analogue is portable and quite flexible in terms of the number of subjects who may participate, the replenishment rates that may be used, and the initial size of the pool. Its limitations are those of the experimenter. As additional subjects are added, the tasks of recording subjects' behavior, pool level, and response rates and following the replenishment schedule become very complex. Some of these difficulties were overcome in an analogue developed by Harper (1978). In Harper's analogue, subjects were simply shown the number of points available in the pool by an overhead projector. On each trial subjects were asked how many points—0, 1, 2, or 3—he or she wished to take. The pool was replenished by doubling on a random schedule averaging every other trial, with a maximum of 15 points. Harper's version of a social trap differed from the other three analogues in that it required subjects to respond on a trial-by-trial basis, whereas the others allowed continuous responding. Both possibilities are amply represented in naturally occurring social traps. Subjects in Harper's analogue behaved quite similarly to subjects in the other analogues. Depletion of the resource pool occurred regularly; communication enhanced the ability of subjects to prolong the life of the pool, to harvest more points, and to allow more points to be replenished. By varying the instructions to subjects, Harper also demonstrated that a cooperative set led to better use of the pool than a competitive set.

Social trap analogues can be developed using a variety of different materials that vary in the sophistication of technology employed and the amount of effort required on the part of the experimenter. In general, a negative correlation exists between these two aspects. Analogues developed with a sophisticated technology require little effort from the experimenter once the electromechanical or computer programming has been accomplished. Analogues that use simple and portable materials usually require a great deal of effort from the experimenter during each experimental session. Within the general structure of the social trap analogue, specific modifications may be made to approximate more closely the features of the situation to which a researcher may wish to generalize. The way in which the resource units are labeled, the manner in which they may be acquired, the replenishment rate, the way in which resource units may be exchanged for goods or services—all these can be varied to approximate more closely a specific situation. Thus, a basic social trap analogue may be used to conduct research designed to explore the kinds of behavior that emerge as a function of a wide variety of independent variables chosen for their theoretical importance. Other analogues may be designed to provide information relevant to specific problems of interest to the experimenter or to policymakers.

OTHER COMMONS DILEMMA STUDIES

Three empirical studies of the commons dilemma are reported here. Two were laboratory studies and one, a field observation.

Stern (1976) built a commons dilemma around the theme of a simulated automobile carpool. A brief report of the basic model for this study is presented in Stern (1977). Over a series of trials, subjects in groups of four had to chose between driving alone or in a carpool. A salary was given at the beginning of each trial, and the costs of driving were subtracted. Subjects choosing to carpool were further required to determine a driver who received an additional payment. The three independent variables studied were: (1) gas allocation by rationing or by prices increasing as a function of consumption; (2) influence attempt: no attempt, positive incentives, or spot messages; and (3) amount of information given about timing of price increases. The results indicated better regulation of the gasoline resource with pricing than with rationing. The difference between these two emerged late in the trials. The spot messages and positive incentives were ineffective. Full information on the timing of price increases led those groups to maintain the resource longer but had only a weak effect toward inducing subjects to select the carpool option.

Stern's study was interesting because it used a realistic task and allowed many different manipulations. The data have important implications for policy decisions. Finding spot messages ineffective was not surprising. We found a similar incident in an interview with the chief of campus police at Arizona State University. The discussion concerned the problem of people parking bicycles around the entrances of the student union and other major buildings. The short-term advantages for bikers were being closer to the goal and having their bikes handy upon exit. As more and more people parked their bikes around the entrances, they became very cluttered and hazardous. Trying to exit a building often required weaving through the maze of bikes. The police chief said they had repeatedly tried placing announcements warning of the danger in the campus newspaper but found that the only effective means of diminishing the problem was confiscating bikes and writing citations.

Another commons dilemma was created in two studies by Dawes, McTavish, and Shaklee (1977). A clever one-trial dilemma was devised. Subjects in groups of four each had to mark an *O* or *X* on a card. People choosing *O* were paid $2.50, with no one getting fined. An *X* paid $12.00, with a $1.50 fine for everybody in the group including the person marking *X*. Thus, large personal gains could be achieved for a solitary defector choosing *X*; but if more than two people defected, group members started losing money.

Independent variables were communication and whether subjects could lose money. Four communication conditions were run: no communication, 10 minutes of communication about an irrelevant topic, 10 minutes of conversation relevant to the decision task, or 10 minutes of relevant discussion followed by a nonbinding vote of their intentions. The loss variable was that subjects either could lose money, or losses would be truncated at zero. In actuality, subjects in the loss condition never lost money except in the session itself. Subjects came to the experiment with three friends. Their winnings or losses were pooled and split evenly. One person's losses were made up by their friends' earnings.

The results indicated an effect only for the communication variable. The least proportion of defections occurred in the relevant-communication condition. The added ability to vote did not increase cooperation.

The second study by Dawes et al. (1977) replicated the communication effect and investigated Kelley and Stahelski's (1970) triangle hypothesis that a competitive person playing against a cooperative person forces the cooperative one to compete. The finding that communication enhances cooperation is consistent with game literature (Wichman, 1972) and other social trap papers (Brechner, 1977; Edney & Harper, 1978).

The last study to be reviewed (Wilson, 1977) is one of the most interesting. Although the experimental controls are weak in places, it is the only test of the commons dilemma in a field setting. Wilson identified two distinct territorial practices among Maine lobster fishermen. One form was similar to the commons in that fishermen from a particular harbor shared an area in an uncontrolled manner. Outsiders were excluded from the area, but relatives and former residents could readily begin fishing if they wished. The second type of territoriality was controlled areas where, through voluntary agreement, the fishermen maintained control over the total number of fishermen in the area. Fishing behavior in the uncontrolled areas took the form of often random placement of traps and placement of traps in nonoptimal areas in order to reserve that space for when temperature changes made it more productive. The controlled-area fishermen were more selective about trap placement, seeking the prime areas.

Wilson had 15 fishermen in both controlled and uncontrolled areas record data on their lobster hauls over three seasons. The data indicated that the lobsters caught in the controlled areas were larger. More were caught and more caught per haul. The females were more likely to be mature. Fishermen in the controlled areas had an average income over $6000 higher than fishermen in the uncontrolled areas. These data were consistent with Cass's (1975/1976) study showing that dividing a common resource into private properties led to better maintenance of the resource.

SOCIAL TRAP ANALOGUES AND OTHER PARADIGMS

The research paradigm most closely related to social trap analogues is the *n*-person game paradigm (Edney & Harper, 1978). *N*-person games have been used sporadically by social psychologists to attempt to generalize to a broader range of situations than is possible with the standard two-person prisoner's dilemma game. In some instances, n-person games have been used simply to highlight the differences between dyads and larger groups (Bixenstine, Levitt, & Wilson, 1966; Kahan, 1973; Weil, 1966). N-person games have also been used for the study of coalition formation (Murnighan, Komorita, & Szwajkowski, 1977). However, there has been little accumulation of closely related experiments in the n-person game literature. Instead, individual investigators have devised games to suit a specific research question. As a consequence, no theory of behavior in n-person games has been developed, just as little theory has been developed to guide research in prisoner's dilemma games (Pruitt & Kimmel, 1977).

The social trap analogues designed by Brechner, Cass, Harper, and Shippee differ from n-person prisoner's dilemma games and most other n-person games in several important ways. First, whereas both paradigms share the feature that a choice that is positive for the individual is negative for the group, in the social trap, the group-negative consequence is delayed and long-term, and the individual-positive consequence is immediate. In n-person games, an individual defecting response has immediate positive consequences for that person and immediate negative consequences for the remainder of the group. This fundamental difference has the consequence that in the social trap analogue, an individual's specific response may have little immediate impact on the other's outcomes, unlike in the n-person game. Thus, subjects are virtually independent for short-term outcomes but are highly interdependent with regard to the long-term outcome of the preservation or depletion of the resource pool. A second way social trap analogues differ from n-person games is that in an n-person game, the valence of cooperative and defecting responses remains constant over the repeated trials of the game; whereas in the social trap analogue, the valence of the various options may shift continuously. A third difference is that social trap analogues allow continuous responding, as in Brechner's, Cass's, and Shippee's work, rather than strictly trial-by-trial responding. Also, there need be far less constraint on the behavior of subjects in a social trap analogue than in an n-person game where typically only two responses are permitted. A range of choices can be given in a discrete trial version of the social trap analogue, whereas in a continuous response version, subjects may respond as slowly or rapidly as they wish. Expanding response options even further, options such as time-outs may easily be incorporated into the procedure.

Thus, the social trap analogue models a rather different and a larger variety of social dilemmas than the n-person game. Social trap analogues address some of the same issues that have been examined by the use of n-person prisoner's dilemma and related games. However, social trap analogues cannot be subsumed under any of these other games. Recent formal analyses of n-person games (Dawes, 1975; Komorita, 1976) do not capture the structure of the social trap or the social trap analogue.

Because of the variation in the behavior of subjects that is allowed within the social trap analogue, because of the wide variety of independent variables that can be manipulated within the paradigm, and because of the range of dependent variables that can be generated, social trap analogues will be useful tools for the analysis of behavior in social interdependence situations. The effects of communication (Brechner, 1977; Harper, 1978), leadership structure, and decision-making rules (Shippee, 1978) have already been studied within the paradigm. Whether or not an investigation is concerned with the social trap analogue as a model of the commons dilemma, energy consumption, or individual self-control, this paradigm offers social psychologists a new setting for research on both basic social processes and applied problems.

SOCIAL TRAPS AND ENERGY

Social traps have a direct relationship to energy in both fundamental and applied ways. On the fundamental level, this section analyzes the reinforcement situation of social traps as a function of the flow of energy through individual and social systems. Then, on the applied level, several examples of social traps that relate to energy use and the use of natural resources are presented.

On a fundamental level, behavior in a social trap is primarily a function of the reinforcement structure. The definition of *reinforcement* is circular; you cannot tell whether something is reinforcing until after you have delivered it and measured its influence on response rates (cf. Catania, 1968, p. 344; Platt, 1973, p. 642). People vary in whether they find a particular environmental stimulus or event reinforcing. Some people like candy, some people do not. In trying to find common aspects of various reinforcers, one is drawn to the underlying notion of energy. The primary reinforcers, such as food and water, are all primary energy sources for the human system. Generalized conditioned reinforcers, such as money, tokens, or food stamps, can all be exchanged directly for energy sources. Social reinforcers, such as smiling or verbal praise, serve to indicate situations where positive reinforcers could be, will be, or will continue to be delivered. Schedules of reinforcement, such as those that approximate the reward structure of natural social traps, can be thought of as schedules of energy distribution where energy is defined as anything that aids a system to produce heat (or movement).

Like that of *reinforcement,* the definition of *energy* is itself circular. Energy is measured in regard to the work it did. Energy often is defined as the ability to do work (Oppenheimer & Borchers, 1963, p. 16; Phillipson, 1966, p. 1). Energy has been defined as the ability to produce heat (Odum & Odum, 1976, p. 266). Einstein (1961) defined energy E of a body m moving at a velocity v that takes up energy E_0 in the form of radiation as

$$E = \frac{mc^2 + E_0}{1 - \frac{v^2}{c^2}}$$

where c is the speed of light. For a body at rest in respect to a reference point, which does not absorb energy, $E = mc^2$. Energy can take many forms: potential (or stored) energy, heat, light, chemical energy, electrical energy, and nuclear energy (Faires, 1970; Wilson, 1963). Likewise, reinforcers also can take those forms.

Because the concepts of reinforcer and energy seem to be related, it could prove useful to analyze the reinforcement situation of social traps as a function of the flow of energy. When energy (or matter) flows, it can set up natural cycles (Morowitz, 1968). For example, when the wind blows, it may pick up a paper plate or a tin can and roll it across the ground. The flow of wind can turn a windmill or cause a hanging sign to swing. The flow of water in rivers produces eddies and whirlpools, or it can turn a turbine. When the sun shines on the earth, temperature gradients are set up between the poles and the equator that create cycles in the movements of air masses and transoceanic currents. When money flows from an employer, it causes workers to return to the workplace every Monday at 8:00 A.M. In social traps, when reinforcers are available, people repeatedly respond for them.

Odum (1971) and Odum and Odum (1976) have devised a symbolic language for analyzing the flow of energy through a system. A *system* is defined as anything that functions as a whole by the integration of organized parts (Odum & Odum, 1976, p. 15). Many different systems can be defined merely by changing the component parts that belong to the system. Figure 2.2 shows an energy-flow analysis of the flow of water in a hypothetical agricultural community. Starting with the sun, the flow of energy can be seen as it transmutes into its various forms and finally emerges on the farms in the form of crops. Each farm system tries to maintain itself and grow. The system that can maximize its energy inputs and use the power must efficiently will survive (Lottka, 1922; Odum & Odum, 1976). From the point of view of these individual consumers, it would be best to have unlimited access to the storage reservoir of water. However, if survival of the larger ecosystem is taken as the focus of attention, it is best to maintain the water storage in the reservoir at high levels.

The social trap becomes clear when we see that each system has its own energy requirements for survival. On an individual level, it is best to have a large flow of water available. On an ecosystem level, it is best to conserve the water and use it at about the rate that it replenishes itself (i.e., cycles). If each farmer acts to maximize short-term production and draws large amounts from the reservoir, the resource will eventually be depleted. The short-term positive consequences that control behavior in social traps generally apply to energy flow in the smaller, individual systems. The long-term negative consequences affect energy flow throughout the larger ecosystem. The long-term consequences, although they eventually come to bear on the individual consumers, have large-scale systemwide effects. A good example can be seen in the behavior of a small manufacturing plant in Los Angeles with which we are familiar. The business

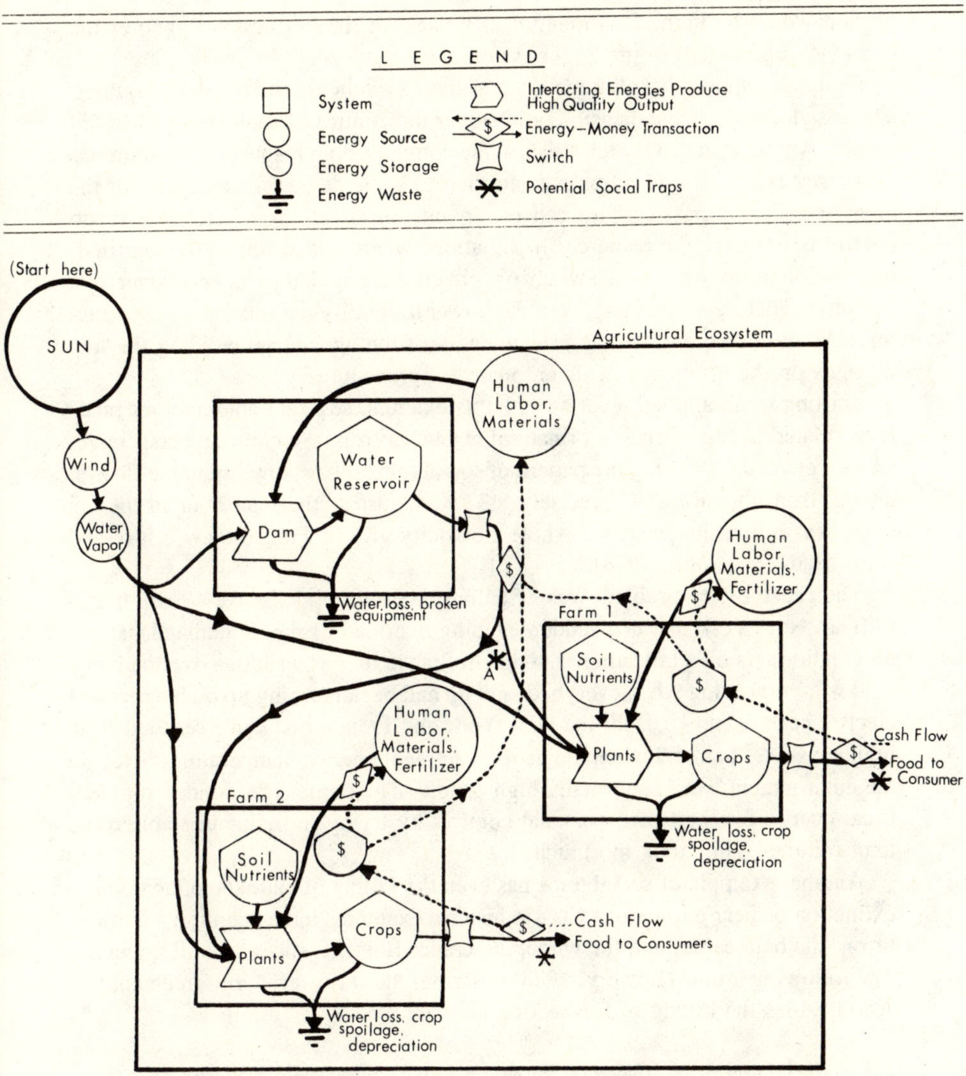

FIG. 2.2. Energy-flow analysis of a hypothetical irrigation system. The social trap at point A is described in the text. (Adapted from Odum (1971) and Odum and Odum (1976)).

uses large quantities of both cold water and steam. During the recent California water shortage, the business responded by purposefully increasing its volume use of water and delaying the installation of a recycling unit. The reasoning was that if the government decided to impose quotas based on normal use rates, their portion would be high. This maximizes power for the business but ignores the energy-flow problem in the larger system.

Figure 2.2 shows that all three systems have switches at their system's output. These switches indicate decision points over the control of each system's stored matter. Anytime such a switch exists at the common-pool resource, the commons no longer exists. The resource is controlled to some degree. The tragedy of the commons in the classic sense can occur only in situations where little or no control exists over the resource. In situations where social traps are identified, one option is to wire in a switch by creating a regulatory agency over that resource. These agencies are given the power to modify the reinforcement structure. As discussed earlier, the effectiveness of the agency in avoiding the trap depends on the efficiency of its enforcement procedures.

Shifting to an applied level now, let us examine several contemporary problems related to use of energy or natural resources from a social trap perspective. One recent event showing the pattern of social traps relates directly to the flow of energy from the sun: "At precisely 8:27 A.M. just as the rush hour in the big cities was hitting its peak, the whole electricity grid of France blew a fuse and went dead. . . . [Cook, 1978]."

The cause of the problem was essentially the same [as the New York blackout]. In New York it was a sudden evening overload of power demand, largely air conditioners on a hot summer night. In France, it was a sudden overload on a cold winter morning when everybody got up and began turning up oil burners and electric heaters [p. 4]. Both the New York and French blackouts resulted from high consumer demand when the general ambient energy (temperature) levels in the environment were abnormally high or low. The humans responded for electrical energy for their own personal comfort by trying to maintain stable room temperatures. They took too much.

Another example of social traps has been the killing of animals to the point of extinction or near extinction. Classic historic examples have been the American bison, the bald eagle, and the whooping crane. But the process is still going on. The following article (Lowery, 1978) describes the short-term reinforcement that accompanies the killing of fur-bearing animals:

> The biggest money makers at the sale were lynx pelts, sold for an average price of more than $400, with one pelt bringing $650. There were 210 lynx pelts sold altogether.
>
> Other furs sold included: 930 beaver, averaging $30 a pelt; 800 mink averaging $30; 300 muskrat at $4 each; 121 otter averaging $90, with the top otter pelt going for $150.
>
> The biggest earning of the sale went to trapper Enoch Flett who collected $5,200 [p. 1].

These short-term benefits of cash for pelts are controlling the trappers while the overall populations of these animals slowly decrease. The long-term results are large-scale imbalances of the energy distribution systems and food chains within the ecosystem. Many people derive all or a portion of their livelihoods from trapping in the North. Any measures taken to protect the populations of fur-bearing animals also must take into account alternative income sources for the trappers, such as reinforcing behaviors incompatible with trapping.

It is interesting at this point to digress and describe how the fur trappers caught the lynx, because the process of animals being trapped is basically the same as the process of humans being caught in social traps. The animal is attracted to one aspect of the environment (bait—short-term rewards) and ignores another (trap—long-term consequences). Sometimes the bait can be rather unusual: "Willie Dysart, with a straight face, told fellow trappers, who had just finished extolling the virtue of fish livers, beaver innards and raw eggs for bait, that what really seemed to attract the lynx was Chanel No. 5, the perfume of Catherine Deneuve" (Lowery, 1978, p. 1).

It is also interesting to note the effects of attempts to protect endangered species. When an endangered species comes under the control of a regulatory agency, the pattern of killing often changes from one of a free commons to one of periodic, time-based slaughter. In Rhodesia's Wankie Game Park ("Elephant orphans," 1978):

> Four hundred of the Park's 12,000 elephants were shot in July and August to keep the population stable. . . .
>
> It is a massive exercise in wholesale slaughter, with herds being picked out by spotter planes conducting a systematic search of the area and by men with high powered rifles on the ground.
>
> "In all our national parks, we have a problem of maintaining the ecosystems," [Dr. David] Cumming, [chief research officer], said. "The parks have only a certain carrying capacity for animals" [p. 3).

The human population problem also can be conceived to be a social trap. Describing all the short-term positive consequences associated with making love is beyond the scope of this paper. The long-term negative consequences could be disease, possibly unwanted children, and a population expanding to the point at which resource production and storage no longer can support all the bodies. On a systemwide level, this relates to energy in that each new child born increases the energy requirements of the society into which the child was born. The introduction of contraceptive devices was a technological advance that changed the reinforcement structure. The probability of some of the shorter-term negative outcomes, such as venereal disease, was reduced; and the long-term negative consequence, overpopulation, could be avoided. If we can believe what we hear about the sexual revolution, the frequency of responding has increased, but the social trap has been avoided because technology has changed the nature of the long-term consequence.

Another area that may qualify as a social trap is the increasing use of automation in industry. As described earlier, the system that maximizes its energy inputs and uses that energy most efficiently tends to survive. Generally in business, it is wise to try to increase the efficiency and accuracy of the business's work procedures. This reduces costs, which increases the profits and helps the business system survive relative to its competitors. Mechanizing procedures is one way to increase efficiency and lower costs, because for many tasks, machines can do the procedures more quickly and accurately than human beings can. In the process, however, jobs are eliminated. For example, the Okazaki plant of Mitsubishi Motors in Japan, which makes Plymouth and Dodge cars for import to the United States, has recently installed computer-controlled robots to weld together the automobile bodies. Luckett (1978) describes:

> It's an eerie sight to watch. Fully articulated arms snake out of their niches from four sides, find their way under and around obstructions, clamp precisely at a predetermined spot—and Zap, it's welded. . . .
>
> "We found that we need fewer welds using robots than when the welds are made manually," my engineer guide told me. "The robot never forgets, or makes a faulty weld, or skips a spot. In one case where we scheduled 15 spot welds by hand our tests showed that nine robot welds gave us a stronger bond." [p. 89].

The robots were installed because they were more accurate and consistent in the short term than human welders. This resulted in a long-term positive gain of reduced costs for the motor company. But hidden among these benefits is another consequence that may not be so beneficial—less work for humans. The company is reducing costs, but in the long run, it is also eliminating customers. The same magazine that contained this story also showed an automated control system being tested by another company that would eliminate the need for test-car drivers. In a disturbing book, *Ecophysics,* Wesley (1974) describes in great detail how increased computerization could lead to a greatly reduced human population in the same manner that oxen and plowhorses were replaced by tractors. The basic idea to be gained from this example is that whenever changes are made by a business, the short- and long-term consequences of the act must be taken into consideration—not only the consequences for the business itself but also for other systems of which it might be a part.

CONCLUDING REMARKS

This chapter has attempted to describe how many situations can be viewed as social traps. A growing body of research has examined social traps in both laboratory and applied settings. Several different tasks have been developed to simulate social traps. Communication was an important variable for maintaining a resource. Dividing the resource into territories over which individuals had control resulted in better regulation than free access to the commons. Feedback was found to be important in order to regulate the resource pool.

A disturbing aspect of the dilemma is whether, as the population increases, commons can be maintained without controls. Hardin (1968) considered the problem one with no technical solution. Crowe (1969) argued that scientific solutions are possible. The basic question is whether the laissez faire attitude of Adam Smith is viable for our future survival. The data from several experiments suggest other alternatives besides governmental regulation. If strong channels of communication and feedback can be developed within a framework of voluntary control, perhaps some of the spirit of Adam Smith's ''invisible hand'' can be preserved. A good example is provided by the fishermen in Wilson's (1977) controlled area who were able to manage a common resource through voluntary agreement. The results of Shippee's social trap analogue suggest a similar solution.

In addition to future experimentation, social trap analysis may prove useful as a training tool for future psychologists, environmentalists, and the lay public. By teaching the principles of social trap analysis or by placing individuals in social trap simulations, the learners can become more sensitive to recognizing situations where long-term consequences could be detrimental. Recently, Powers, Duus, and Norton (1979) have been using a commons game based on the prisoner's dilemma to teach students about social traps. But awareness of the long-term consequences is only half the solution. Changing behavior controlled by the short-term consequences is the other half. Hopefully, social trap analysis can help provide some of the answers.

ACKNOWLEDGMENTS

Completion of this chapter was greatly facilitated by a sabbatical leave granted to the second author by Arizona State University. During that leave, facilities and assistance were graciously provided by the Department of Psychology at the University of Colorado—Boulder.

REFERENCES

Ainslie, G. Specious reward: A behavioral theory of impulsiveness and impulse control. *Psychological Bulletin,* 1975, *82,* 463-496.

Bixenstine, V. E., Levitt, C. A., & Wilson, K. V. Collaborations among six persons in a prisoner's dilemma game. *Conflict Resolution,* 1966, *10,* 488-495.

Brechner, K. C. An experimental analysis of social traps: A laboratory analog. (Doctoral dissertation, Arizona State University, 1974). *Dissertation Abstracts International,* 1975, *35/11,* 5658B. (University Microfilms No. 75-10, 037)

Brechner, K. C. An experimental analysis of social traps. *Journal of Experimental Social Psychology,* 1977, *13,* 552-564.

Brechner, K. C. *Superimposed schedules, social traps, and the prediction and control of behavior.* Unpublished manuscript, 1978. (Available from the author, 107 South Fairoaks, Pasadena, Calif. 91105.)

Bureaucracy buries 1-man coal mine. *Arizona Republic,* June 12, 1972, p. 23.

Cass, R. C. Territoriality and the tragedy of the commons: A social trap analysis (Doctoral dissertation, Arizona State University, 1975). *Dissertation Abstracts International,* 1976, *36/06,* 2996B. (University Microfilms No. 75-28, 482)

Catania, A. C. (Ed.). *Contemporary research in operant behavior.* New York, Foresman, 1968.

Cook, D. Rush hour power failure blacks out chilly France. *Los Angeles Times,* December 20, 1978, Pt. 1, p. 4.

Crowe, B. The tragedy of the commons revisited. *Science,* 1969, *166,* 1103-1107.

Dawes, R. M. Formal models of dilemmas in social decision-making. In M. Kaplan & S. Schwartz (Eds.), *Human judgment and decision processes.* New York: Academic Press, 1975.

Dawes, R. M., McTavish, J., & Shaklee, H. Behavior, communication, and assumptions about other people's behavior in a commons dilemma situation. *Journal of Personality and Social Psychology,* 1977, *35,* 1-11.

Deutsch, M., & Krauss, R. M. The effect of threat upon interpersonal bargaining. *Journal of Abnormal and Social Psychology,* 1960, *61,* 181-189.

Deutsch, M., & Krauss, R. M. Studies of interpersonal bargaining. *Journal of Conflict Resolution,* 1962, *6,* 52-76.

Edney, J. J. The nuts game: A concise commons dilemma analog. *Environmental Pychology and Nonverbal Behavior,* 1979, *3,* 252-254.

Edney, J. J., & Harper, C. The commons dilemma: A review of contributions from psychology. *Environmental Management,* 1978, *2,* 491-507.

Einstein, A. *Relativity.* New York: Crown Publishers, 1961.

Elephant orphans frolic at reserve. *Los Angeles Times,* December 20, 1978, Pt. 1-A, p. 3.

Faires, V. M. *Thermodynamics* (5th ed.). New York: MacMillan, 1970.

Hardin, G. J. The tragedy of the commons. *Science,* 1968, *162,* 1243-1248.

Harper, C. *The role of communication and cooperative or individualistic orientation in a simulated resource management dilemma.* Unpublished master's thesis, 1978, Department of Psychology, Arizona State University.

Kahan, J. P. Noninteraction in an anonymous three-person prisoner's dilemma game. *Behavioral Science,* 1973, *18,* 124-127.

Kelleher, R. T. Conditioned reinforcement in second order schedules. *Journal of the Experimental Analysis of Behavior,* 1966, *9,* 475-486.

Kelley, H. H., & Stahelski, A. V. Social interaction basis of cooperators' and competitors' beliefs about others. *Journal of Personality and Social Psychology,* 1970, *16,* 66-91.

Komorita, S. S. A model of the n-person dilemma type game. *Journal of Experimental Social Psychology,* 1976, *12,* 357-373.

Lottka, A. J. contributions to the energetics of evolution. *Proceedings of the National Academy of Science,* 1922, *8,* 147-155.

Lowery, B. Perfume attracts the beast. *Winnipeg Free Press,* December 21, 1978, p. 1.

Luckett, H. P. PS/What's news. *Popular Science,* 1978, *213* (4), 89.

Marr, M. J. Second-order schedules. In D. P. Hendry (Ed.), *Conditioned reinforcement.* Homewood, Ill.: Dorsey Press, 1969.

Morowitz, H. J. *Energy flow in biology.* New York: Academic Press, 1968.

Murnighan, J. K., Komorita, S. S., & Szwajkowski, E. Theories of coalition formation and the effects of reference groups. *Journal of Experimental Social Psychology,* 1977, *13,* 166-181.

Odum, H. T. *Environment, power and society.* New York: Wiley, 1971.

Odum, H. T., & Odum, E. C. *Energy basis for man and nature.* New York: McGraw-Hill, 1976.

Oppenheimer, S. L., & Borchers, J. P. *Direct and alternating current.* New York: McGraw-Hill, 1963.

Phillipson, J. *Ecological energetics.* London: Edward Arnold, 1966.

Platt, J. Social traps. American Psychologist, 1973, *28,* 641–651.

Powers, R. B., Duus, R. E., & Norton, R. S. *The commons game: Teaching students about social traps.* Paper presented at the meeting of the Rocky Mountain Psychological Association, Las Vegas, 1979.

Pruitt, D. G., & Kimmel, M. J. Twenty years of experimental gaming: Critique, synthesis and suggestions for the future. In M. R. Rosenzweig & L. W. Porter (Eds.), *Annual review of psychology* (Vol. 28, pp. 363–392). Palo Alto, Calif.: Annual Reviews, 1977.

Rachlin, H. Self control. *Behaviorism,* 1974, *2,* 94–107.

Shippee, G. *Leadership, group participation, and avoiding the tragedy of the commons.* Unpublished doctoral dissertation, 1978, Department of Psychology, Arizona State University.

Solnick, J. V. *Self control and human choice behavior.* Paper presented at the meeting of the American Psychological Association, Toronto, 1978.

Stern, P. C. Effect of incentives and education on resource conservation decisions in the simulated commons dilemma. *Journal of Personality and Social Psychology,* 1976, *34,* 1285–1292.

Stern, P. C. Psychological research and the steady state society. *Population and Environmental Psychology Newsletter,* 1977, *4,* 10–15.

Thomas, R. L. Hopis charge Navajos overgraze joint-use range. *Arizona Republic,* June 12, 1972, p. 23.

Weil, R. L. The n-person prisoner's dilemma: Some theory and a computer-oriented approach. *Behavioral Science,* 1966, *11,* 227–233.

Weiss, B., & Laties, V. G. Fractional escape and avoidance on a titration schedule. *Science,* 1958, *128,* 1575–1576.

Weiss, B., & Laties, V. G. Titration behavior on various fractional escape programs. *Journal of the Experimental Analysis of Behavior,* 1959, *2,* 227–248.

Weiss, B., & Laties, V. G. Characteristics of aversive thresholds measured by a titration schedule. *Journal of the Experimental Analysis of Behavior,* 1963, *6,* 563–572.

Wesley, J. P. *Ecophysics.* Springfield, Ill.: Charles C Thomas, 1974.

Wichman, H. Effect of communication on cooperation in a 2-person game. In L. Wrightsman, J. O'Conner, & N. Baker (Eds.), *Cooperation and competition: Readings on mixed-motive games.* Belmont, Calif.: Brooks/Cole, 1972.

Wilson, J. A. A test of the tragedy of the commons. In G. Hardin & J. Baden (Eds.), *Managing the commons.* San Francisco: Freéman, 1977.

Wilson, M. *Energy.* New York: Time-Life Books, 1963.

Wrightsman, L. S., Jr., O'Conner, J., & Baker, N. J. *Cooperation and competition: Readings on mixed-motive games.* Belmont, Calif.: Brooks/Cole, 1972.

3 Applications of Behavioral Analysis for Conserving Transportation Energy

David A. Reichel
E. Scott Geller
Virginia Polytechnic Institute and State University

The application of behavioral analysis to transportation management has significant potential for lessening the impact of current transportation behaviors on the energy crisis. Such application also holds potential for solving problems of traffic congestion and pollution, and could enhance transportation planners' abilities to analyze and predict transportation behaviors and to maximize the utility of transportation systems. This chapter reviews the research designs, procedures, results, and conclusions of behavioral interventions aimed at conserving transportation energy.

ENVIRONMENTAL IMPACT OF THE AUTOMOBILE

A guiding principle for selecting targets for energy conservation measures should be the fact that the bigger energy consumers can become the bigger energy conservers. The largest consumer of transportation energy in the United States is the private automobile. Indeed, 57.1% of all energy consumed for transportation, more than 9 trillion BTUs, has been attributed to urban and inner-city auto trips. Further, as much as 34.1% of the total vehicle miles traveled each year is traveled as work-trips, consuming 22,386 million gallons of fuel (Andrle & Dueker, 1974). Thus the reduced or more efficient use of the automobile is a socially valid objective for behavioral interventions.

In addition to the tremendous amount of energy that the automobile consumes, there are other adverse environmental impacts as well. For example, in 1966, before technological innovations such as catalytic converters and unleaded gasoline, the automobile contributed up to 60% of the air pollution in the United States (Dorf, 1974). The environmental impact of noise and land pollution (e.g., traffic jams with honking horns and automobile graveyards) are other negative aspects of the automobile.

In terms of social-environmental impact, the automobile has changed the shape of our cities (Dorf, 1974). Indeed, Los Angeles, California, has often been referred to as "the city the automobile built." The crowded, congested streets and traffic jams that are experienced in so many cities today are counter to the prime advantage of the automobile—that is, a personalized and flexible mode of transportation. Most tragic, of course, is that more than 40,000 deaths each year result from automobile accidents (Dorf, 1974).

The economic costs of the automobile are staggering. Dorf (1974) noted that the average car consumes 700 gallons of gasoline a year. At "only" $1.20 per gallon, this amounts to more than $800 per year per car for fuel; and with 100 million cars in the United States, the amount spent on gasoline reaches astronomical levels. Additionally, $20 billion is spent on additions and repairs to roads annually, and up to two-thirds of the land in urban areas is devoted to freeways, gas stations, parking lots, and other auto-related uses (Dorf, 1974). Perhaps the most infamous fact about the automobile is that in terms of energy efficiency (calories/gram/kilometer plotted against body weight), it is less efficient than a man on a bicycle, a horse, and a jet transport and is roughly as efficient as a pigeon (Dorf, 1974).

THE "SOCIAL TRAP" OF TRANSPORTATION ENERGY USE

Stern and Kirkpatrick (1977) noted that energy consumption may be viewed as a "social trap." Specifically, energy consumption is a behavior that results in immediate benefits but long-term costs. For example, use of the private automobile results in the immediately pleasurable consequences of convenient, flexible mobility. However, the long-term unpleasant consequences stemming from increased demand include reduced gasoline supplies, higher prices, and the other negative environmental impacts already noted.

Because immediate consequences control behavior, transportation energy consumption is readily rewarded. Further, the delayed unpleasant consequences of consuming transportation energy are distributed across the population, and punishment is "shared." Usually, individuals can not be expected to forego individual rewards in order to avoid a share of delayed punishment (Stern & Kirkpatrick, 1977). Stern (1979) argued that "the problem of conservation is to get people to judge energy use by its social consequences and to gauge their behavior accordingly [p. 16]." As this chapter clearly demonstrates, behavioral approaches have relied almost exclusively on external, individual incentives.

However, one conclusion that is drawn herein supports Stern's notion that new social norms are needed. Stern and Kirkpatrick (1977) found new social norms to develop in laboratory simulation games that involved energy conservation decisions. Perhaps group, rather than individual, reinforcement contingencies are more apt to encourage social norms in the real world (cf. Reichel & Geller, 1979; 1980).

IDENTIFYING TRANSPORTATION CONTINGENCIES

Potential Consequences of Driving and Not Driving

Everett and Hayward (1974) proposed a speculative analysis of the contingencies associated with two mutually exclusive behaviors—car-driving and not-car-driving. The 2 × 2 contingency table in Table 3.1 summarizes this analysis. Assuming that the goal is to facilitate the use of transportation modes other than car-driving, Cell 3 represents the most favored approach. A reduction of the consequences in Cell 1 would entail a reduction of reinforcing events, a politically difficult manipulation that is questionable on ethical grounds. An increase

TABLE 3.1
Speculative Analysis of Consequences of Car-Driving and Not-Car-Driving[a]

	Consequences of Behavior	
	Pleasant	*Unpleasant*
car-driving	1. short travel time high prestige choice of route choice of departure/arrival time deferred costs privacy large cargo capacity	2. traffic congestion high gas and maintenance costs
not-car-driving	3. freedom from driving responsibilities healthful exercise from walking and biking	4. long travel time low prestige minimal choice of travel route inflexible travel schedule immediate costs crowding inconvenient cargo transport dangerous (walking, biking)

[a]Adapted from Everett and Hayward (1974, Fig. 1).

in Cell 2 consequences is theoretically inappropriate, as the behavior that would be negatively reinforced by the avoidance of these aversive consequences would not be controllable. However, most alternatives to car-driving are perhaps preferable, and the specification of the not-car-driving behaviors is not vital. Further, the theoretical inappropriateness of such an approach should neither discourage nor be used as a rationale against applied research involving aversive consequences. Interestingly, this approach has been favored by the Carter administration. The National Energy Plan proposed a graduated excise tax on "gas-guzzlers" and a standby gasoline tax to take effect if total national gasoline consumption exceeded annual targets (President Carter, 1977). Such taxes could be viewed as incentives to adopt alternative transportation modes, such as carpooling, but the results of this review suggest that increased cost may not serve such a function (cf. Andrle & Dueker, 1974; Hirst, 1976).

Cell 4 manipulations again leave unspecified the behaviors to be reinforced, but as already noted, this may not be critical. According to Everett and Hayward (1974), Cell 3 is the most favorable approach, not so much as a result of a process of elimination but for its political and behavioral control advantages. In fact, this strategy has been rather extensively applied by Everett and his colleagues (e.g., Deslauriers & Everett, 1977; Everett, 1973; Everett, Hayward, & Meyers, 1974) and is reviewed in a later section of this chapter.

Referring to Table 3.1, it is noteworthy that deferred cost (e.g., credit purchases of gasoline) is speculated upon as a reinforcing consequence of car-driving, whereas immediate cost is an aversive consequence of not-car-driving. Hayward and Everett (1976) attempted to make the cost of car-driving salient to the drivers in an effort to reduce miles driven in their private automobiles. Ninety solo drivers were divided into four feedback groups: (1) mileage feedback only (control); (2) operating cost feedback at 7¢ per mile; (3) operating and depreciation cost feedback at 15¢ per mile; and (4) operating, depreciation, and social cost feedback at 25¢ per mile. The social costs were arbitrarily determined in an attempt to reflect costs due to air pollution, health decrements, and defacement of the environment. Each subject received daily feedback on the windshield of his or her car for a period of 2 weeks. During treatment, miles driven per day were compared to pre- and post-intervention baselines (ABA design). Participants' attitudes toward the automobile as a viable mode of transportation were assessed before and after the feedback interventions.

Intragroup and intergroup comparisons showed *no* mileage changes induced by the various feedback manipulations. Hayward and Everett (1976) proposed two factors that may account for the failure. First, real response consequences were not manipulated, because subjects most likely still paid through a credit system; that is, delayed costs were not made immediate. Second, because the work-trips were mandatory and no viable alternative transportation systems were available, it would have been difficult (and very inconvenient) for the participants to reduce their mileage. However, it could be argued that awareness of cost was made immediate, and this may account for the more negative attitudes that participants held of the car as a viable mode of transportation after cost feedback compared to attitudes before cost feedback. Specifically, there were stronger changes (i.e., a larger move toward more negative attitudes) for groups with the higher cost feedback (e.g., the 25¢-per-mile group). This result raises two issues—attitude-discrepant behavior, which is addressed in a later section, and the present issue of delayed versus immediate monetary costs.

In a market research study to design a public transportation system, Golob, Canty, Gustafson, and Vitt (1972) had 1603 interviewees from the Detroit area rate a number of design, service, and convenience factors using paired-comparison and semantic differential techniques. The system was called the Demand-responsive Jitney (D-J) and proposed door-to-door service with user-shared cost. Counter to Everett and Hayward's (1974) speculations concerning costs, subjects rated cash at boarding as the most favored method of fare collection and credit cards as the least favored on both scaling techniques. The implication is that immediate cost may not always be the most undesirable. It may be that under conditions where transportation is mandatory (e.g., the work-trip), the monetary cost of transportation, or at least the time interval between a given trip and the payment for that trip, does not determine one's selection of transportation mode. Thus the identification of salient contingencies in transportation behavior is an area that needs intensive investigation; the role of response cost is one example where experimental validation is still lacking.

The contingency table in Table 3.1 does raise a number of critical comments. First, the reader can probably suggest a number of additional consequences that could be added to the matrix. For example, in Cell 2, the fact that the automobile is sometimes a risky alternative could be added. The consequences of safety, monetary savings, and energy conservation could be added to Cell 3. Moreover, it could be argued that the inclusion of "dangerous" in Cell 4 may not be appropriate, comparing actual safety records of automobiles with those for other transportation modes (e.g., walking, busing, and even biking), although it may be that most individuals perceive the automobile as the safest mode of travel, especially when they are driving (i.e., in perceived control).

Secondly, as a model, the matrix in Table 3.1 is probably too simplistic. The unpleasant consequences listed in Cell 4 and the pleasant consequences listed in Cell 3 presumably result from not using the automobile. However, the pleasant and unpleasant consequences of alternatives such as walking, bus riding, biking, and even carpooling or vanpooling are not the same. The matrix fails to differentiate the consequences among these alternatives. For example, "minimal choice of travel route" may be an unpleasant consequence of bus riding, but it is generally not relevant to walking and biking. Indeed walking may provide more travel routes than driving a car. Similarly, "crowding," "inflexible travel schedule," and "immediate costs" are not relevant to walking and biking and are less relevant to carpooling and vanpooling than to bus riding. Further, the current national popularity of physical activities such as jogging and biking raises doubt concerning "low prestige" as an unpleasant consequence for walking and biking.

Gaming Simulations

It is evident that the contingencies surrounding mode of transportation selection are yet to be clearly defined and experimentally validated. Yet definition and validation could become quite costly in terms of time and money. Gaming simulation is one research approach to studying transportation systems that is less costly than actual large-scale experimental manipulation and validation. Everett, Studer, and Douglas (in press) noted three gaming simulations that dealt with transportation. Specifically, the Urban Systems Model (Environmetrics Inc., 1972) concerned decision making with respect to auto expenses, mass transit routes, and setting of transit fares; the Transportation Planning Simulation (Applied Decision Systems Inc., 1970) simulated citizen participation in transportation planning during a transportation crisis; and Transportation Policy Negotiation, or TRANSPON (Hargroves & McGray, 1973), demonstrated the interactions and problems involved in the planning of transportation facilities.

Everett et al. (in press) noted that none of these simulation games manipulated variables according to behavioral technology and then described a game that did apply behavioral contingencies. Groups of two or three subjects were assigned to play the game for 1 hour. They were told to pretend that they had just moved into a new city and that for each of the 10 rounds of the game, they were to select a mode of transportation for going to work and for making shopping trips. Subjects could choose a private car, a carpool, a bicycle, a bus, or walking. Each choice would cost the players varying amounts of dollars, free time, and ease-of-living points. At the beginning of each round, players were given cost–payoff tables that listed the costs for the various selections and were then given their "income." Players then calculated their costs and payoffs. There were five groups in the study: (1) baseline, with no consequences for selecting the bus for either shopping or commuting; (2) continuous reinforcement, where each player received $20 for choosing the bus for commuting and another $20 for choosing the bus for shopping; (3) variable-ratio 6 (VR6), where each player threw a die when choosing the bus to commute and threw a die when choosing the bus to shop, and then received $20 if the number 4 appeared; (4) variable-ratio 36 (VR36), where the player threw two dice when choosing the bus to commute and two dice when choosing the bus to shop, and received $20 if the number 12 was rolled; and (5) an ABA design, whereby players received the baseline condition for the first three rounds, then VR6 for the next four rounds, and were then returned to the baseline condition for the last three rounds.

The results of this gaming simulation revealed that low rates of bus riding occurred under baseline, that the VR6 condition led to the highest level of bus selection, and that the continuous-reinforcement and VR36 conditions led to moderate levels of bus riding. These results are quite analogous to those obtained with rats as subjects (cf. Ferster & Skinner, 1957). Specifically, less dense schedules of reinforcement often promote higher response rates than does continuous reinforcement.

If the results of such gaming simulation continue to duplicate results obtained in other laboratory manipulations, their use may expedite the identification and validation of the contingencies that influence selections of transportation mode. Gaming simulation is relatively inexpensive and could be used to "pretest" costly and untried manipulations in a variety of settings. However, as Everett et al. (in press) noted, the results of simulations cannot be equated with actual behavioral observation. In other words, gaming simulation may be used to develop and refine potential interventions, but actual large-scale manipulation of the contingencies is required before objective conclusions about an intervention can be made. Indeed, gaming alternatives may not exist in reality. For example, the monetary "credits" received for selecting the bus in the game are not typical consequences of actual bus riding. Further, the credits may not be comparable to the monetary savings incurred by actually riding a bus.

STRATEGIES FOR TRANSPORTATION MANAGEMENT

Behavioral interventions to conserve transportation energy have been directed toward three categories of transportation behaviors: (1) driver habits, (2) ride sharing, and (3) public transportation. Energy-related driver habits may be dichotomized as related to two specific dependent variables: vehicle-miles-of-travel (VMT) and miles per gallon (mpg). Ride sharing involves three targets: carpooling, vanpooling, and ad hoc pooling. Interventions with public transportation have usually targeted bus riding.

Reducing Vehicle-Miles-of-Travel (VMT)

Reducing VMT is perhaps the most direct approach toward reducing automobile usage. Indeed, if each of the 100 million automobiles in the United States were driven 10 miles less per week, 3.5 billion gallons of gasoline, or 5% of the annual automobile demand for gasoline, would be saved annually (Federal Energy Administration, 1977). Foxx and Hake (1977) published the first successful application of behavioral technology to reduce directly the use of one's automobile. Two groups were used: A reward group received the opportunity to earn various prizes for reducing automobile mileage, whereas a control group received no such incentives for reducing mileage. The 12 subjects in the reward group were enrolled in a course taught by one of the experimenters, whereas the

9 subjects in the control group were enrolled in a course taught by an instructor not affiliated with the study.

An ABA design was followed. No contingencies were in effect during the initial baseline. Subjects were required to bring their cars to an odometer checkpoint at designated times. On the final day of baseline, subjects in the reward group were given a guide that listed mileage-reduction goals from 10% to 50%, the prizes associated with each reduction, and the miles-per-day reduction needed to achieve the reductions. Prizes were awarded in $5 increments for each 10% reduction in mileage relative to initial baseline. Further, a tour of a mental health facility was offered to subjects with a 20% or greater reduction. Subjects with the two greatest percent reductions were refunded the price of their university parking stickers ($10), and the person achieving the third greatest reduction received a free oil change and lubrication. Baseline conditions continued for the control group during this phase of the study.

During the 28-day intervention period, odometer checks were made for each subject in the reward and the control conditions. Odometer checks were made at the regular checkpoint or by requesting subjects to indicate on a map where his or her car was parked and having recorders go to the car and read the odometer. These odometer checks also served to ensure that the subjects had driven their own cars. On the final day of interventions, subjects were told that the experiment was over and that prizes would be awarded on the final day of class. Subjects who had not qualified for a prize and all subjects in the control group were told that $1.50 would be paid for attending all regular odometer checks. On the last day of follow-up, a final odometer check was made for all subjects, and prizes were awarded when subjects brought their cars to the checkpoint.

A mean reduction of 20% from initial baseline was obtained by subjects in the reward condition. Subjects in the control group showed a mean increase in mileage of 4.5% from initial baseline. During follow-up, subjects in the reward condition increased their mileage to a level equivalent to that of initial baseline. Subjects in the control condition maintained their previous level.

A number of procedural weaknesses in this study were corrected in a replication and extension by Hake and Foxx (1978). In the second study, a short-term reinforcement schedule was incorporated by providing small weekly reinforcers in addition to the larger, end-of-treatment reinforcers of the earlier study. All subjects were commuting students, eliminating one problem of the earlier study, which included dormitory residents with minimal essential driving. Because subjects in the reward group of the earlier study (Foxx & Hake, 1977) were students in one of the experimenter's classes, a design artifact could have been responsible for the obtained mileage reductions. Hake and Foxx (1978) examined the behavioral effects of this potential artifact (which they termed a "leader" variable).

Specifically, one-half of the subjects in the reward condition were from a university class taught by the second author ($N = 4$), and one-half were from a class whose instructor was not involved in the study ($N = 5$). All the subjects in a self-recording condition ($N = 8$) were from the same class taught by the second author. During the 3rd week of initial baseline, subjects were matched between reward and control conditions on the basis of number of miles driven. During initial baseline, students were told that they were part of a study on driving habits and would eventually be able to earn money for their participation. Such notice before intervention presumably reduced experimental demand effects (e.g., a Hawthorne effect) that could have resulted from the intervention. On the final day of baseline, subjects in both groups were given guides describing the experimental contingencies for their particular conditions.

During treatment, subjects in the reward condition could earn from $1 to $2 per week for 10% to 30% weekly mileage reductions relative to initial baseline. As in the previous study, the guide given on the last day of baseline showed each subject the miles-per-day reduction needed to achieve the 10% to 30% reductions. Additionally, subjects could earn $5 for averaging a 10% mileage reduction over the 28 days of intervention, with $2.50 increments for each additional 10% reduction up to 30%. On the last day of the final baseline (following the 4-week intervention period), $10 each was awarded to the two drivers with the greatest mileage reductions relative to initial baseline. Lastly, there was a $2.50 attendance prize for stopping at a checkpoint before leaving the campus on the 8 odometer-check days during treatment and on each of the 4 checks during final baseline. All these contingencies were listed in the guide that was given to the subjects on the final day of initial baseline. As in the previous study, "map checks" were also used to record mileage and ensure that subjects were driving their own cars.

For the self-recording group, subjects could earn $1.25 each week for calculating their average miles driven per day for at least 5 of the 7 weekdays. A monthly prize of $6 was awarded to subjects who calculated their mileage for at least 5 days per week in each of the 4 weeks of treatment. Additionally, there were two $10 awards for the two most accurate records and $2.50 prizes for attendance at all regular odometer checks. As with the reward group, these contingencies were listed in the guide given to subjects on the last day of initial baseline. Final baseline was instituted following 28 days of intervention, during which all contingencies were removed except for the $2.50 attendance prize for arrival at the regular odometer checks.

The reward group obtained a 22.5% reduction in miles driven relative to initial baseline. The leader variable, manipulated only under the reward condition, did indeed have a strong effect. The four students from the second author's class averaged a 40.3% reduction, whereas the five students from the other class averaged only an 8.2% reduction. The self-recording group obtained a 12.6% reduction in miles driven relative to initial baseline. All subjects in this condition were students of the second author's class; thus the reduction may have been biased by the leader variable.

These results clearly demonstrated that contingency management may be applied effectively to reduce automobile mileage. Furthermore, the "leader variable" is an interesting phenomenon that demands further study, because it may be applied very cost-effectively to reduce automobile mileage. The effect of a leader in the Hake and Foxx (1978) study suggests that supervisors, managers, and foremen in organizational settings could be incorporated into mileage-reduction programs to motivate individual participation and to facilitate the efficacy of a self-recording intervention. The observed effect of a leader in the Hake and Foxx study could have been due to several factors—among them being a special respect for the author and his intentions, the prompt that students received by seeing the author in class, and possibly special statements made by the author. Obviously, further research concerning this "social attention phenomenon" is indicated.

A cost–benefit analysis by Hake and Foxx was based on absolute reduction in mileage relative to the average of both baselines and revealed a total of $79.97 savings in terms of gasoline not purchased by the reward group; whereas for the self-recording group, an overall savings of $20.40 was obtained. Total prize money was $126.00 and $109.50 for the reward and self-recording groups, respectively. Although such an analysis does reflect more costs than savings, this type of analysis can be criticized for not realistically reflecting the cost–benefit exchange. The total $100.37 saved in terms of gasoline not purchased by subjects is an "elusive" benefit that does not represent a tangible accruement to subjects or to experimenters. Hake and Foxx (1978) recommended that the procedures would be more economical under any of the following conditions: (1) if the size of the monetary prizes were reduced; (2) if only drivers who use their cars frequently were participants; and (3) if prizes were awarded for absolute miles reduced per day instead of for percentage reduction from baseline. This latter suggestion would eliminate the problem that a large percentage reduction may not necessarily result in an economical reduction if the subject drives very little.

A procedure that may be more economical and just as motivating is the use of a lottery or raffle. Geller and his students (Geller, Chaffee, & Ingram, 1975; Ingram & Geller, 1975; Witmer & Geller, 1976) have demonstrated the effectiveness of a raffle in encouraging participation in paper-recycling programs. One "chance" in a lottery or raffle could be made available for each 10% reduction in mileage per week, with additional bonus "chances" made available for each 10% reduction achieved over a month.

Reichel and Geller (1980) successfully employed public feedback and a contest to reduce VMT. There were two specific goals of the study. First, the potential enhancement of other strategies suggested by the leader variable was examined within the context of an organizational setting. Specifically, participants were involved in an undergraduate research program at Virginia Polytechnic Institute and State University under the supervision of this chapter's second author. All participants had been involved in this program for a minimum of 3 months. It was assumed that this situation was similar to work groups that may be found in industrial and business settings. The 11 participants were divided into two teams. Team A ($N = 5$) consisted of 4 undergraduates and a secretary. Team B ($N = 6$) consisted of 5 undergraduates and Geller.

The second goal was to reduce VMT with a cost-effective procedure (i.e., a contest combined with public feedback). Specifically, participants were given verbal instructions and a flyer describing the contest. The team that achieved the greatest percentage reduction in average VMT reduction over the contest period relative to baseline would receive various prizes that had been donated by local merchants. The prizes ranged in value from approximately $5 to $10. The teams were instructed that the winning team could distribute the prizes among its members in any way desired.

Throughout all phases of the experiment, each participant recorded his or her daily odometer reading at approximately the same time each day. Additionally, participants were paired and made daily odometer readings of their partner's cars to obtain reliability estimates. Further, the experimenter made spot checks of the participants' odometers.

An ABA design was used. Baseline lasted 4 weeks. Participants were instructed on how to read the odometer and how to record the reading on the data sheet. On the final day of baseline, participants were divided into the two teams and informed of the contest. On the 1st day of the contest, the total weekly VMT and average daily VMT for each participant and for each team for the last 2 weeks of baseline were posted in the research office.

The contest lasted 3 weeks. At the end of each week, the total weekly VMT, the average daily VMT, and the percentage change from baseline for individuals and for teams were posted. At the end of the 3rd week, the winning team was declared and the contest contingency removed. Follow-up recordings continued for 1 week, cut short by the end of classes.

The average VMTs for the last 2 weeks of baseline were 918.5 miles and 1419.3 miles for Teams A and B, respectively. The average reduction in VMT for the 3 weeks of the contest were 208.1 miles and 881.7 miles for Teams A and B, respectively. These average reductions in VMT corresponded to percentage reductions of 22.7% and 62.1%, respectively. The VMTs for the 1 week of follow-up were 1376.2 miles and 565.7 miles for Teams A and B. These corresponded to a percentage increase from initial baseline of 49.8% for Team A and a decrease of 60.1% for Team B.

The team with the "leader" achieved a percentage reduction nearly three times that of the other team; thus the potential enhancement of a strategy by a leader was again demonstrated. The use of the contest combined with public feedback demonstrated a strategy that can be applied cost-effectively. The application of group contingencies may be one way of encouraging the development of new social norms (cf. Stern, 1979; Stern & Kirkpatrick, 1977). In addition, the participation of leaders or supervisors may be important for such development.

Encouraging Fuel-Saving Driving

Lauridsen (1977) reported a successful effort to reduce gasoline consumption in automobiles through feedback and lottery interventions. Four employees of a utility company who typically drove over 50 miles a day, and often more than 100 miles a day, served as subjects. A fuel-flow meter was installed in each of four company vehicles driven by the four subjects. In two of the automobiles (Drivers 1 and 2), the meter provided continuous miles-per-gallon (mpg) feedback. In the other two vehicles (Drivers 3 and 4), the meter provided continuous gallons-per-hour feedback. Prior to baseline the meters were installed, and each driver was informed of the meter and its feedback function.

During baseline the meters were covered and rigged so that removal of the cover could be detected. The drivers were asked to drive as they normally did. Mean mpg was recorded daily for all four vehicles by one observer and at least once a week by a second observer. During initial feedback the meter was uncovered so that the drivers could receive feedback. Mean mpg was again recorded daily. Drivers 1, 2, and 3 were then returned to baseline conditions, followed by 2 days of second feedback. Following the 2 days of initial feedback for Driver 4 and the 2 days of second feedback for Drivers 1, 2, and 3, a 5-day feedback-plus-lottery (FL) condition was implemented. In addition to feedback from the meters, all drivers had a daily opportunity to earn a chance for a lottery. Each driver earned one lottery ticket for every .1-gallon increase over his individual median mpg for the last 5 days of the preceding baseline (i.e., second baseline for Drivers 1, 2, and 3 and initial baseline for Driver 4). Additionally, a driver lost one ticket for every .1-gallon decrease in mpg. At the end of each day, the driver was informed of this mpg for the day, the number of chances earned or lost for the day, and his cumulative number of chances earned or lost. The lottery winner received the total estimated dollar savings (based on projected gasoline saved) for all four drivers (i.e., $20.49). Following the FL condition, all drivers were returned to baseline conditions. A second 5-day FL condition was then implemented in which only Drivers 1 and 3 were able to participate. The winner received $8.07.

During initial feedback, Drivers 1, 2, and 4 obtained increases in mpg of 7.92%, 6.66%, and 15.18%, respectively. Driver 3 showed a decrease in mpg of 9.67%. During second feedback, Drivers 1 and 2 obtained increases in mpg relative to second baseline of 24.44% and 30.50%, respectively. Driver 3 showed no change relative to second baseline, and Driver 4 did not participate in this phase. During the first FL condition, Drivers 1, 2, 3, and 4 obtained increases in mpg relative to initial baseline of 7.92%, 6.66%, 5.64%, and 5.06%, respectively. In the second FL condition, Driver 1 obtained an increase in mpg of 6.93% relative to initial baseline, and Driver 3 showed a decrease in mpg of 7.25% relative to initial baseline. Drivers 2 and 4 did not participate. Lauridsen (1977) suggested that the failure to increase mpg for Driver 3 during feedback and FL might have been due to unrepresentative data during initial baseline, which varied from 40 to 338 miles a day and from 9.7 to 14.9 mpg.

The consistent increases obtained by Drivers 1 and 2 when feedback was available suggests that mpg feedback might have been more effective than gallons-per-hour feedback. The relative efficiency of feedback and FL was not clarified. Interviews with the drivers indicated that when feedback was available, they tended to accelerate more slowly, maintain constant speeds, avoid streets with frequent stops, release the accelerator when going downhill, and reduce idling time. All these techniques are considered to increase fuel efficiency (cf. American Automobile Association, 1975; Federal Energy Administration, 1977).

Two comments should be made concerning the Lauridsen (1977) study. First, a separate lottery condition without feedback would have served to identify the relative efficacy of the lottery contingency independent of feedback. Second, an interesting question is raised in that the drivers adopted well-established fuel-conserving behaviors when feedback was available. It is not clear whether these behaviors arose from the feedback itself or from previous knowledge about fuel-conserving driving behaviors. A suggestion that the change in driving habits may not have been due entirely to the feedback meters is that because of the nature of the internal combustion engine, there may be a temporal delay between any given driving behavior, its effect on the engine, and its effect being manifested on a feedback meter (Hurley, 1978). Further research is indicated, for increasing fuel economy may be a viable alternative to reducing miles traveled, especially for mandatory trips such as commuting to and from work.

Runnion, Watson, and McWhorter (1978) reduced the amount of fuel consumed by a textile company's distribution fleet by using feedback, social recognition, and a lottery. The participants were 195 long-line and intermill drivers. The researchers studied mpg as the dependent measure, which was calculated daily for intermill drivers (Experiment 1) and on a per-trip basis for long-line drivers (Experiment 2). For Experiment 1, initial baseline lasted 1 week. During this time the intermill drivers received no encouragement to increase mpg. The first intervention lasted 1 year and consisted of the following: (1) antecedent instructions concerning ways to increase mpg and an announcement of a program to increase mpg among the drivers; (2) personal recognition, on a random basis, in the form of letters commending the driver's improved performance; (3) public feedback of mpg for the fleet and for each individual, which was graphed and posted daily in the drivers' room at each of the 58 terminals; (4) social reinforcement from company officials in the form of nonsystematic and unstructured praise for improved performance; and (5) peer competition prompted by public display of individual performance.

During the first 2 weeks of the 2nd year, the second intervention of Experiment 1 was implemented and consisted of the following: (1) publicly displayed feedback of fleet performance; (2) the social reinforcers obtained during the 1st year; (3) a personal letter to each driver informing him of his past performance; (4) public feedback of each driver's previous day's performance; and (5) a weekly lottery for pen sets, key chains, and the like with the company logo. For the lottery, a driver's name was added to the pool for each day of the week he obtained above 6 mpg. The lottery and personal letters were implemented on an unspecified intermittent schedule.

During baseline the average mpg for the intermill drivers was 5.73. For the 1st year the average was 6.02 mpg, a 5.06% increase above baseline. The average for the 2nd year was 5.97 mpg, a 4.18% increase above baseline.

In Experiment 2 the same contingencies implemented during the 1st year of Experiment 1 were applied to the long-line drivers. The 2nd year consisted of the 2nd-year contingencies used in Experiment 1 plus public feedback for each long-line driver team. During baseline the long-line fleet averaged 4.80 mpg. The 1st-year average was 4.90 mpg, a 2.08% increase above baseline. The average for the 2nd year was 5.23 mpg, an 8.96% increase above baseline.

In terms of energy savings, the textile company saved enough fuel to run its entire fleet for 1 month at no cost. Two design limitations noted by Runnion et al. were: (1) a short baseline period; and (2) no control groups, nor a reversal condition or a multiple-baseline comparison. These limitations were largely the result of the company's desire to institute money-saving procedures immediately. No cost–benefit analysis comparing the dollars saved in gasoline not purchased to the cost of implementing the contingencies was reported. The Runnion et al. (1978) study did extend the findings of Hake and Foxx (1978) by using a non-college-student population and by applying seemingly low-cost, perhaps even cost-efficient, contingencies.

Ride Sharing

Another approach to reducing automobile usage has focused on ride-sharing programs such as carpools and vanpools. Ad hoc pooling, or hitchhiking, is a third form of ride sharing that has received some attention (Barkow, 1974; Dallmeyer, 1975). Attempts to increase carpooling have employed two basic techniques: (1) preferential treatment of carpoolers (Letzkus & Scharfe, 1975; MacCalden & Davis, 1972; Rose & Hinds, 1976); and (2) organizational efforts for carpools (Andrle & Dueker, 1974; Jones & Derby, 1976; Scheiner & Keiper, 1976). A number of employers have combined these two techniques (cf. Pratsch, 1975; United States Department of Transportation, 1975). Other researchers have evaluated the influence of attitudes and psychosocial factors on the formation and maintenance of carpools (Barkow, 1974; Dueker & Levin, 1976; Horowitz & Sheth, 1977).

Employer-sponsored vanpools have increased significantly in the past 5 years. Currently, at least 33 employers operate vanpools that provide employees with an economical means of getting to and from work (cf. Forstater & Twomey, 1976). A typical eight-passenger vanpool saves approximately 5000 gallons of gasoline a year (Pratsch, 1977). Bent (1977) estimated that if 3100 vanpools were in operation by 1985, 1.218 million vehicle-miles-of-travel would be saved, representing an annual savings of 19 million gallons of gasoline. Furthermore, carpool programs could result in an annual savings of from 1 to 3 billion gallons of gasoline (United States Department of Transportation, 1975).

Priority Lanes. Physical interventions such as priority lanes for carpoolers are one of the more commonly used techniques to promote carpooling. Priority lanes offer users the incentives of decreased travel time and avoidance of traffic congestion. A number of communities have implemented priority lanes for carpools, including Dade County, Florida (Rose & Hinds, 1976); Boston, Massachusetts (Miller & Deuser, 1976); Fairfax County, Virginia (Fleischman, 1975); and San Francisco, California (MacCalden & Davis, 1972).

MacCalden and Davis (1972) reported an attempt to increase carpooling in the San Francisco Bay area. Commuters who carpooled to San Francisco (i.e., at least three riders per car) received reductions in monthly fares and the use of a faster priority lane on the San Francisco–Oakland Bay Bridge. There was some increase in the use of carpools. However, enforcement of the contingencies was a problem in that the priority lane was located in the center of the bridge, making virtually impossible the apprehension of lane users who were not carpooling. Additionally, some people tried to undermine such contingencies by placing mannequins in their cars to give the appearance of riders (Ricc, 1975).

The problem of enforcement is critical to the success of priority-lane incentives. Miller and Deuser (1976) noted that priority lanes that are physically separated from general-use lanes by safety cones or barrier walls do not pose a substantial enforcement problem. On the other hand, priority lanes that are not physically separated from general-use lanes, such as those in Dade County, Florida, have posed substantial enforcement problems. The principal problem is that of lane use by noncarpoolers. Conventional patrol techniques such as high visibility (i.e., presence of many patrol cars) and apprehension of violators have been almost singly relied upon to enforce priority-lane restrictions. For example, in Dade County, six police officers patrolled the 5.5-mile priority lane when the restrictions were in force (Rose & Hinds, 1976). But six police officers for 5.5 miles of highway does not seem cost-effective, especially when there is still an 8% violation rate. Innovations in enforcement techniques for priority lanes are needed (Miller & Deuser, 1976).

In addition to the enforcement problem accompanying priority lanes, the strategy that some solo drivers have employed to meet the contingencies may reduce the overall energy conservation potential of a community transportation program. Specifically, Brasted (1978) observed that on the Shirley Highway, in Fairfax, Virginia, which has priority lanes for both buses and carpoolers, drivers pick up riders at bus stops so as to qualify for the carpool lane. Such practices reduce the number of passengers carried by bus, thereby increasing the energy intensiveness of the bus system (i.e., the ratio of fuel used to passenger miles).

Rose and Hinds (1976) reported on the Dade County priority-lane project. A contraflow bus lane and a carpool lane were designated for 5.5 miles of the South Dixie Highway, which links suburban South Dade County with the Miami central business district. During morning and evening rush hours, the two inside lanes were reserved for buses or cars with two or more occupants. Overhead message boards above the lanes notified drivers of the lane restrictions. Results showed a fivefold increase in bus ridership levels over the level that existed before the project was implemented. However, at the time of implementation, the number of peak-period bus trips was increased from 10 to 84, making it difficult to specify the separate impact of the priority lane on bus ridership. Survey results showed that 65.1% of the bus riders were former solo drivers and that 12.5% had previously ridden in a carpool. Of the 740 bus riders surveyed, 94% verbalized that they used the bus to get to and from work; 49.1% reported that one reason for changing to the bus was low cost (50¢ to 60¢); 47.4% reported improved work-trip travel time (a 15- to 20-minute decrease) as a reason for changing; and 77.4% reported convenience as a reason. Vehicle and vehicle occupancy counts showed that the carpool priority lane, along with a carpool parking lot in the central business district, increased the average occupancy per vehicle for all general-use lanes from 1.30 to 1.45 persons per vehicle during the initial 6 months of the project. The carpool lane accounted for 40% of the rush-hour person-trips during the project hours (i.e., 7 to 9 A.M. and 4 to 6 P.M.). Fuel savings were reported as a reason for changing to a carpool by 62.6% of surveyed carpoolers, time advantage by 58.7%, and companionship by 37.4%. Unfortunately, Rose and Hinds (1976) did not report the previous mode used by carpoolers, nor percentage increases in carpooling.

Preferential Parking. Preferential-parking incentives are perhaps the most frequently used technique to encourage carpooling. For example, the National Aeronautics and Space Administration (NASA) in Houston, Texas, initiated a carpool program in 1964. Presently, 90% of the parking permits at NASA are allocated to carpools, and an average of 3.85 employees ride to work in one car (Pratsch, 1975). Employees from both Jantzen, Inc., and Baltimore Federal agencies were matched for carpools, and the lists were distributed to the employees. The matching efforts increased carpooling by 30% and 7%, respectively. Hallmark Cards and the McDonnell Douglas Corporation employed both matching techniques and parking incentives, and each increased carpooling by approximately 50% (Pratsch, 1975).

Hirst (1976) reviewed the energy impact of carpooling policies in Washington, D.C. Baseline levels of solo drivers, carpools, and mass-transit riders were 52.9%, 25.4%, and 14.5%, respectively. Close-in parking spaces were then reserved for carpoolers, which resulted in a 22.15% increase in carpoolers, a .4% increase in mass-transit riders, and a 10.69% decrease in solo drivers. Following this, a $2-per-day parking charge for solo drivers was implemented along with the reserved parking for carpoolers. There was a 43.82% increase in carpools relative to baseline and a 4.55% increase in mass-transit riders. Further, there was a 22.27% decrease in solo drivers relative to baseline.

Bryant (1974) summarized the incentives that could be offered by employers to encourage carpooling among employees. These incentives included preferential parking, financial bonuses, social recognition, and flexible work schedules to optimize pickup and delivery while avoiding rush-hour congestion. One example of financial rewards was reported by Letzkus and Scharfe (1975). Monetary rewards were effective in increasing carpooling among employees of the United Gas Pipeline Company in Houston, Texas. Employees were paid up to 90¢ a day for using the public bus to get to and from work. Employees who participated in carpools with four or more passengers were allowed free parking at the company's parking garage. Additionally, the company paid two-thirds of the parking fee for three-passenger carpools and one-third of the fee for two-passenger carpools. Thus the incentives not only encouraged carpooling but large carpools as well. Participation during the 1st year included 508 of the 685 company employees. The total cost to the company was $105,900—$74,300 being paid to the employees who rode the bus and $28,080 to the 158 employees who participated in carpools of four or more. Also, there were 15 three-passenger carpools and 12 two-passenger carpools. Unfortunately, no baseline data were reported, nor was a cost–benefit analysis included in order to evaluate the excessive cost of this program.

Organizational Efforts. Another strategy for increasing carpooling involves company organizational efforts. More specifically, information pertaining to the employees' work schedule, address, and willingness to participate in a carpool is collected and usually fed into a computer. The computer program then matches employees for carpools. A list of potential carpools is then returned to the employees. Andrle and Dueker (1974) studied carpool organizational efforts with minimal employer incentives. Three workplaces in Iowa City, Iowa, were studied—a white-collar firm, a blue-collar firm, and the faculty/staff of the University of Iowa. Survey questionnaires, including attitudinal and work-trip items, were administered to employees. Potential carpools were compiled on the basis of responses to the questionnaire and were made available to the employees. Three months after the carpool information had been distributed, a follow-up survey was given to the employees to determine the efficiency of the efforts to organize carpools.

At the University of Iowa, carpoolers were offered second-priority parking (i.e., behind handicapped employees); no incentives were offered by the white- or blue-collar firms. The study was conducted during the 1973–74 energy crisis; thus rising gasoline prices and gas station lines probably provided incentives to carpool. At the white-collar firm, 111 employees indicated a willingness to carpool; but at the 3-month follow-up, only 4 people had actually formed new carpools, and of these 4, 3 had given them up. Similarly, no new carpools were formed at the blue-collar firm. No follow-up survey was conducted for the university employees, although carpool-parking-permit applications showed *no* appreciable increase. Thus without extrinsic employer incentives and no organizational effort beyond the mere distribution of information concerning potential carpools, gasoline prices do not yet appear to provide an incentive for carpooling. Again it appears that cost may not be a salient contingency in one's selection of transportation mode. Indeed the reliance of the Carter administration on aversive consequences to reduce automobile usage (i.e., higher gasoline prices) has not been observed to be a viable method of reducing auto usage. Actually, a gradual price increase in gasoline would allow for adaptation to increasing prices (cf. Stern & Kirkpatrick, 1977).

The results of the attitudinal items in the Andrle and Dueker study reflect the problem of attitude–behavior discrepancies (previously raised by Hayward & Everett, 1976). Across the three firms surveyed, 51.2% of the employees reported a willingness to carpool; yet only four carpools were actually formed. Andrle and Dueker concluded that the potential carpooler must have a fairly regular work schedule, live relatively far away from work, and want to save money. Initially, all potential carpoolers will object to taking more than 15 additional minutes to get to work and will be concerned with the inconvenience of carpooling and the independence sacrificed by joining a carpool. The major reasons for not carpooling that were given in the follow-up evaluation (which consisted of a checklist) were: (1) variable work schedule; (2) no interest; and (3) too much time spent waiting and traveling. Clearly, the inconvenience associated with the carpool is a major deterrent to carpooling. In the initial survey to determine willingness to carpool, 61% of the respondents reported inconvenience as either important or very important. Thus, carpools may be recognized (and reported) as desirable in terms of benefits to society (i.e., energy conserving) but not desirable (nor feasible) in terms of benefits to the individual.

Scheiner and Keiper (1976) reported a carpool-matching effort in Wilkes-Barre, Pennsylvania, that was aimed at three markets: (1) the general public, (2) employers, and (3) labor unions. The general public was approached through media advertising of the carpool-matching effort. Employers were requested to distribute carpool-matching questionnaires to employees with a memorandum that explained the project and endorsed energy conservation. Labor unions were asked to provide names and other information necessary for the matching program. Through these efforts, 2 people were matched for carpools through the areawide approach; 699 employees were matched through the employer approach; and 954 were matched through the labor union approach. This matching occurred in late 1974. In early 1975, 5% of those who had been matched were interviewed over the telephone. Of those surveyed, only 5.9% had formed carpools. Thus it was estimated that only 20 carpools were formed out of the 1655 people who had been matched by the carpool service. Scheiner and Keiper (1976) did not provide an analysis of the number of carpools formed by each market approach. Similar to the findings of Andrle and Dueker (1974), matching efforts alone did not result in a substantial increase in carpooling; it appears that without incentives to carpool, carpool-matching and organization efforts are relatively ineffective.

However, Jones and Derby (1976) did present evidence that contradicts the notion that organizational efforts (without explicit incentives) are relatively ineffective. The city of Sacramento, California, provided a dial-in service designed to provide carpool-matching services for the entire community. This dial-in service was promoted through media advertising and was similar to the areawide approach reported by Scheiner and Keiper (1976). Further, employees were contacted through their employers and were asked to complete and return a carpool-matching questionnaire if they were interested in carpooling. After 6 months, 5083 names had been collected. Approximately 634 of these people were contacted for a follow-up interview. Of those interviewed, 21.8% had been placed in carpools (37% from the dial-in service and 19.3% from the employer service). Of those placed, 15.2% had dropped out (10.9% from the dial-in service and 18.5% from the organizational service). To summarize, of those interviewed, 92 people from the dial-in service had been placed in carpools, and 10 had dropped out; 119 people from the employer service had been placed in carpools, and 22 had dropped out.

Overall, the carpool-matching services reported by Jones and Derby, which were highly similar to those implemented in Wilkes-Barre, were markedly more effective than the Wilkes-Barre and Iowa City efforts. As in the Wilkes-Barre and Iowa City efforts, the only general incentive to carpool in the Sacramento project was the gasoline savings the individual could realize. The times at which the projects were implemented did not differ; thus the most obvious explanation for the difference in these efforts to form carpools is the different populations represented. Iowa City is a relatively small college town; Wilkes-Barre is a medium-sized urban area where approximately 40% of the work force is employed in manufacturing (Scheiner & Keiper, 1976); whereas Sacramento is a large city with many state and federal employers.

Such matching efforts could perhaps be facilitated if individuals were supplied not only with a list of potential carpools but with tips on how to form a carpool as well. Such tips could include how to decide who will drive, who will be picked up first, how to handle gasoline purchases, and so forth. However, research conducted by Jacobs, Fairbanks, Poche, and Bailey (1979) suggests that the inclusion of such tips does not increase the effectiveness of matching programs.

Table 3.2 summarizes the effectiveness of selected, employer-sponsored efforts to increase carpooling. The results suggest a tentative conclusion that incentives may have more impact than matching efforts; but both together may be optimal. This notion is partially supported by the increased carpooling obtained where no matching efforts had occurred (i.e., Rose & Hinds, 1976). However, Bryant (1974) noted that even though there is no known correlation between being matched for a carpool and subsequent participation in a carpool, inadequate research may be responsible for the low correlation. The typical problems of field research and case study evaluations suggest only cautious acceptance of this conclusion. For example, the lack of control groups, the inability to control other variables such as increasing gasoline prices, and frequent failures to report baseline carpooling limit most research on the effectiveness of employer-sponsored carpool programs.

Jacobs, Fairbanks, Poche, and Bailey (1979) used token reinforcement and reserved parking to increase carpooling among students at a large university. Four parking lots served as data collection sites; two (Lots A and B) served as

TABLE 3.2
Effectiveness of Selected, Employer-Sponsored Carpooling Programs[a]

Employer	*Pretreatment percentage of Carpooling Employees*	*Increase in Percentage of Carpooling Employees During Treatment*
	Matching Service Only	
Air Force Accounting and Finance Center	25	20
Baltimore Federal agencies	–	7
Bell Helicopter	20	20
Boeing Aircraft	29	8
Jantzen, Inc.	–	30
Northern Natural Gas	32	27
Tennessee Valley Authority	33	14
Texaco	8	14
Mean	24.5%	17.5%
	Matching Service and/or Priority Parking	
Federal Highway Administration	–	55
Hallmark Cards	–	50
McDonnell Douglas	–	50
Washington, D.C., area	–	22
Mean		44.25%

[a]Sources: Pratsch (1975); United States Department of Transportation (1975, p. 23).
Note: – indicates that no data were reported.

treatment lots and two (Lots C and D) as control lots. Students were required to pay $12 for a parking sticker in order to use the lots. Further, Control-Lot C had a mechanical gate that required 25¢ to operate.

During the 9 days of baseline, the number of cars entering each lot, the number of occupants per car, and the license plate number of each car were recorded. Reliability readings were made on one-half of the 40 days during which the experiment was conducted. During the 6th to 8th days of baseline, drivers entering Lots A, B, and C received a flyer that listed the benefits of carpooling, indicated the location of parking spaces that would be reserved for carpools, and requested occupants to complete a computer-match form.

During treatment, approximately 45 spaces each in Lots A and B were marked "Carpools Required from 7:30–10:00 A.M." Each carpool was met at the reserved parking entrance by an observer who explained the reserved parking contingencies. In addition, each person in a carpool (i.e., in a car with two or more occupants) received a token redeemable for 25¢ in merchandise at local campus stores. Students who had completed and returned the computer-match forms were sent a list of students who lived within one-half mile of them and who were interested in carpooling. Additionally, a flyer was included that listed tips on forming a carpool and a graph of how much could be saved by carpooling. On the 6th day of treatment, a flyer reminding people of the contingencies was placed on all cars in Lots A, B, and C. On the 20th day of treatment, cars in Lot A were notified that the tokens would no longer be distributed but that reserved parking would still be available.

In Lot A, carpooling increased from an average of 12% during baseline to 20% during the combined reserved-parking/token phase of treatment. When reserved-parking-only was in effect, carpooling averaged 22%. In Lot B, carpooling averaged 11% during baseline and 14% during treatment. In Lot C, carpooling declined from 11% to 9.8%; and in Lot D, carpooling declined from 9.4% to 8.9%.

A cost-benefit analysis was performed that compared the cost of program implementation with the amount of money that could be saved by requiring less parking spaces to accommodate the same number of people. The results of this analysis showed that the program cost $1135, whereas the savings in terms of parking spaces was $454,000 (i.e., 15 spaces saved at $3600 per space).

Jacobs et al. (1979) suggested that the differences in carpooling increases between Lots A and B may have been due to the fact that in Lot A, the reserved parking spaces represented 35% of the total spaces available, whereas in Lot B, the reserved parking spaces represented only 8% of the total. Presumably this could have made reserved parking more prominent in Lot A, such that students might have reasoned that if they wanted to park in Lot A, they should seriously consider carpooling.

Perhaps the most revealing findings of this study were the maintenance of carpooling in Lot A after the removal of the token contingency and the fact that interviews with carpoolers revealed that the computer-match list induced very few individuals to carpool. These observations lend further support to the conclusions drawn from studies reviewed earlier (i.e., Andrle & Dueker, 1974; Hirst, 1976; Pratsch, 1975; Scheiner & Keiper, 1976; United States Department of Transportation, 1975). Specifically, preferential treatment alone (i.e., priority parking) may be sufficient to encourage carpooling; computer matching of carpools and incentives such as redeemable tokens may not be necessary. Additionally, scheduling conflicts were reported as the major reason for not carpooling and schedule compatibility, program incentives, and monetary savings as the major reasons for carpooling. These self-report data are consistent with the findings of Andrle and Dueker (1974) reviewed earlier.

Unfortunately, Jacobs et al. (1979) reported no long-term follow-up rates of carpooling. It would be informative to know whether carpools that were formed during treatment were maintained after all contingencies were removed. Perhaps once a carpool has been formed, participants find natural contingencies to maintain the carpool without the necessity of continual incentives (e.g., Barkow, 1974; Dueker & Levin, 1976; Owens & Sever, 1977).

Psychosocial Factors in Carpooling. Sociological and psychological influences on carpooling have been examined by a number of researchers. Barkow (1974) pointed out that carpooling is a social arrangement and that to understand the contingencies involved in promoting carpooling better, it is necessary to develop an understanding of the psychosocial nature of carpooling. For example, Dueker and Levin (1976) compared the attractiveness of hypothetical carpools by having college students rate the relative desirability of hypothetical carpools that varied in number of riders, sex of each rider, and whether or not each rider was an acquaintance. For one-rider carpools, both males and females rated carpools with an acquaintance as more attractive than carpools with a nonacquaintance, and they rated carpools with a female rider as more attractive than carpools with a male rider. The results also yielded a significant interaction between sex of the rider and rider acquaintance. That is, if the rider was an acquaintance, then the sex of the rider had little consequence on the attractiveness ratings; but if the rider was not an acquaintance, then both males and females preferred female riders. Similar trends were obtained for two- and three-rider carpools.

Barkow (1974) concluded from interviews with carpoolers working in the Canadian Ministry of Transportation and Communication that carpooling is for mature people and that the success of a carpool rests on flexibility and mutual accommodation. Resistance to carpooling was seen as stemming from people's hesitancy to encounter and negotiate with strangers. This is not surprising in light of Dueker and Levin's (1976) finding that acquaintances were preferred to strangers. In the Barkow (1974) study, the most frequently mentioned advantages of carpooling were money savings and relief from the tensions of driving, whereas the most frequently mentioned disadvantages were hypothetical constraints on freedom (e.g., obligations to be punctual, reliable, and civil to fellow riders).

Horowitz and Sheth (1977) proposed a mathematical model of ride sharing that was based on socioeconomic level, travel characteristics, and attitudes. The data were collected from questionnaires administered to Chicago-area residents who were contacted through employers. Among the principal conclusions were: (1) Demographic and travel characteristics were poor predictors of driving alone or carpooling; (2) appeals to the issues of energy, traffic congestion, and air pollution had little influence on attitudes toward ride sharing, although respondents from a high socioeconomic level were more sensitive to such public-cost appeals; (3) perceptions toward lost time and inconvenience need to change before travel behavior will change; and (4) in general, solo drivers had a neutral attitude toward ride sharing. These conclusions suggest that certain promotional campaigns based on ride-sharing benefits to the general public (i.e., reduced energy consumption, reduced traffic congestion, and reduced air and noise pollution) will be generally ineffective. On the other hand, it may be effective to aim at alleviating concerns about the loss of independence, loss of time, and loss of convenience associated with ride sharing. For example, time can be gained (not lost) through ride sharing by allowing additional time to read the morning paper, plan the daily agenda, or socialize with fellow riders.

Margolin, Misch, and Dobson (1976) employed a panel discussion technique to generate hypotheses about the incentives and disincentives of ride sharing. Homogeneous consumer groups were used, including groups of (1) central business district, blue-collar commuters; (2) suburban white-collar commuters; (3) suburban executive and professional commuters; and (4) high school drivers. Discussions began with open-ended and general questions and ended with discussion of transportation problems and potential solutions. The results suggested that "driving" connotes different things to different people, and incentives and disincentives vary accordingly. For example, teenagers stated that they experience pleasure and satisfaction from driving as well as a sense of freedom from parents, whereas laborers were apparently much less dependent on driving for satisfaction and pleasure but expressed the need for job-related driving. Thus, to provide incentives for transportation modes other than solo driving, it may be necessary to establish an incentive package that offers a variety of rewards, each geared to some specific population.

The following points summarize the research on carpooling reviewed in this section:

1. Community interventions such as the designation of priority lanes for carpools have been effective in increasing the average occupancy per automobile during peak-period commuter traffic (Rose & Hinds, 1976).

2. The critical issue of enforcement of priority-lane restrictions must be resolved more cost-efficiently.

3. The effectiveness of carpool formation efforts is equivocal; Andrle and Dueker (1974) and Scheiner and Keiper (1976) found carpool-matching services to be relatively ineffective in initiating carpools, whereas Jones and Derby (1976)

reported similar organizational efforts to be relatively effective. Demographic and sociological factors may account for the discrepancy between investigations. Further, it is not clear if additional incentives such as priority lanes are needed to increase the effectiveness of organizational efforts, although incentives alone, without organizational efforts, have facilitated carpool formation (Rose & Hinds, 1976).

4. Some perceptions about carpooling are not influential in the initiation of a carpool but may become rewards for maintenance of a carpool. For example, reduced personal and public costs were not found to be sufficient incentives for initiation of a carpool (Andrle & Dueker, 1974; Horowitz & Sheth, 1977) but may serve as rewards for the maintenance of already-established carpools (Barkow, 1974).

5. The initiation and maintenance of carpooling are moderated by social factors such as sex, prior acquaintance with riders (Dueker & Levin, 1976), and social skills (Barkow, 1974), as well as by demographic and reference-group factors (Margolin et al., 1976).

This final point has important ramifications. Nietzel, Winett, MacDonald, and Davidson (1977) recommended the use of computer matching to form carpools. Indeed, the carpool formation projects reported by Scheiner and Keiper (1976) and by Jones and Derby (1976) used computerized matching services. However, computer-formed carpools are impersonal and do not overcome the reservation people feel when negotiating with strangers (Barkow, 1974). Furthermore, the information needed for computerized formation of carpools runs counter to current concerns about privacy (Margolin et al., 1976).

Overcoming Psychosocial Barriers to Carpooling. If acquaintances (Dueker & Levin, 1976) and social skills (Barkow, 1974) are determining factors in carpool formation and maintenance, then a more personalized approach is necessary. Such an approach could take the form of an employee "carpool contest," whereby employees from specific neighborhoods or work sections would compete for prizes awarded by the employer to those groups with the highest percentage of carpoolers. Alternatively, company-time "carpool parties" could be organized where employers give their employees the opportunity to meet and become acquainted with others interested in forming carpools. Indeed such a procedure was employed by Hewlett Packard Company in Colorado Springs, Colorado (Pratsch, 1975). Computer-matching services could not be practically employed because of variable work schedules. Extended "coffee breaks" were arranged for employees from predetermined residential locations. During these "coffee breaks," employees discussed possible arrival times and routes and were able to negotiate and compromise, overcoming the limitations of a computer-based matching effort. As a result, 40% of the employees participated in carpools. Such techniques afforded the opportunity to develop acquaintances, to generate group cohesiveness, to indicate company commitment to carpooling, and to facilitate a social attention phenomenon similar to that identified by Hake and Foxx (1978) as a determinant of energy-conserving automobile usage.

In addition to techniques aimed at lessening the aversive social factors involved in organizing carpools, the additional time spent traveling in a carpool could be offset by extra time at coffee breaks and lunch, and/or company-car use for carpooling employees when emergencies arise. Additional time allowed at coffee breaks and lunch could even serve as an incentive for carpooling. Again the use of techniques that demonstrate company commitment will take advantage of the social attention or "leader variable" phenomenon (cf. Hake & Foxx, 1978).

Vanpooling. A second major form of ride sharing is the vanpool. The 3M Commute-A-Van system (CAV) reported by Owens and Sever (1977) provides a cost-efficient model for increasing ride-sharing behavior. With continued growth and hiring at the 3M corporate headquarters in St. Paul, Minnesota, the need for additional parking space in a limited area posed a substantial problem in the early 1970s. To avoid investment in a costly new parking facility and to reduce traffic congestion, 3M developed the CAV. The CAV avoided a major problem of carpooling—that is, the requirement of at least one person willing to use his or her own car with compensation barely adequate to meet operating expenses (Owens & Sever, 1977).

For specific residential areas in the St. Paul vicinity, a vanpool coordinator was selected. This coordinator was charged with driving the van, picking up and returning passengers, arranging for service, maintaining the van, training backup drivers, billing riders, keeping records, and parking overnight. An innovative system of contingencies was arranged in order to encourage participation as a coordinator, including a free ride to and from work, personal use of the van at 7¢ per mile, the receipt of all passenger fares over the minimum required (see following), and convenient parking at 3M. Comments from riders surveyed revealed a diverse set of positive consequences for riders, ranging from reduced gasoline and insurance bills and lack of tension from driving to socializing with fellow riders and "sitting next to Mary" (Owens & Sever, 1977, p. 79).

The CAV program currently operates on a self-amortizing basis (i.e., the income generated by the program pays for the cost of the program). Fares are based on operating costs (gasoline and maintenance) and on fixed costs (initial cost, depreciation over 4 years, insurance and licensing). Initially, the monthly fares ranged from \$19.50 to \$29.50 per passenger, depending on the residential area from which the van operated. Fares collected from passengers in excess of eight were given directly to the pool coordinator, providing an incentive for the coordinator to enlist the maximum number of riders (i.e., 11). Table 3.3 summarizes the benefits of the CAV program for 3M users, nonusers, and the general public.

A CAV rider survey conducted in 1976 revealed that 49% of the riders were former solo drivers. Further, there had been no significant decline in carpooling,

TABLE 3.3
Benefits of the 3M Commute-A-Van Program[a]

Benefits to 3M	Reduced congestion at 3M installations Reduced demand for parking Reduced capital expenditures for auto-related facilities More efficient use of land at 3M installations
Benefits to users	Reduced expenses Reduced risks and tensions of commuting Greater availability of car for use by family members, thereby increasing their mobility and opportunities
Benefits to nonusers	Reduced congestion and parking demand in and around the 3M installations
Benefits to general public	Reduced congestion on streets and highways Reduced land use for auto-related facilities Less air and noise pollution Reduced energy consumption

[a] Adapted from Owens and Sever (1977, Table 5, p. 23).

even though 22% of CAV riders formerly participated in a rotating carpool and 16% were former carpool riders. Thus employees (presumably solo drivers) not participating in the CAV program began to carpool. This could be the result of increased awareness of carpooling as a transportation alternative and/or an effort to demonstrate potential for a successful vanpool in an area not presently served by the CAV.

In 1974 a survey was made to compare mode of transportation to work among 3M employees with a similar survey conducted in 1970. The average gross occupancy per vehicle increased from 1.24 in 1970 to 1.55 in 1974, a 25% increase. It is apparent that the CAV directly and indirectly reduced automobile usage. Specifically, from 1970 to 1974, there were 1755 more employees (a 22% increase from 1970) and 98 fewer solo drivers a (1.6% decrease from 1970). Perhaps even more encouraging is that 97% of the CAV riders reported an intention to use the van as their permanent means of getting to and from work (Owens & Sever, 1977).

Ad Hoc Pooling. A third form of ride sharing is what Barkow (1974) referred to as ad hoc pooling. Yukobousky and Fichter (1976) proposed a formalized ad hoc system that centered on the notion of a "mobility club." The two essential components of such a system are: (1) a central dispatcher who matches the travel demands of individuals without an automobile to the trip-making intentions of people with an automobile; and (2) a system to acquaint people in the club efficiently, so that drivers and riders feel comfortable with one another. The individual in need of a ride simply calls into the central dispatcher to be matched with individuals who will be driving to the same area. This system was proposed for small-town, rural transportation needs and, to the best of our knowledge, has not yet been implemented. A third component that may be necessary is reinforcement contingencies to promote and maintain such a system. Thus, more may be needed to attract drivers than the intrinsic rewards of having helped an individual in need of a ride. For example, drivers could be offered reduced parking rates and/or special parking areas.

The notion of ad hoc pooling was extended to formalized hitchhiking by Dallmeyer (1975). This plan legalizes hitchhiking through a voluntary system of registration of hitchhikers and designation of pull-over and pickup points. The City of Eugene, Oregon, was the first to implement a program designed to facilitate hitchhiking. The Eugene Ride Stop program was initiated in 1972 and provided designated spots where anyone in need of a ride could wait for a motorist to stop. The pickup points were located away from intersections and provided pull-over space that did not obstruct traffic, thereby reducing the potential for accidents caused by motorists stopping to pick up a rider (Whitlow & Manzanares, 1973). The program was canceled after approximately 1 year because of neighborhood complaints and increased crimes (e.g., muggings). The pull-over points actually became hunting grounds for criminals looking for victims (Barron, 1978).

A hitchhiking program poses a number of concerns. First, a program designed to promote hitchhiking may hinder mass-transit efforts; that is, bus ridership could decrease if free, low-risk rides were readily available. Second, legal considerations such as accident insurance and liability pose problems. For example, does a city-sanctioned hitchhiking program imply that it is safe to ride in a registered vehicle, and can the city be held liable for injuries incurred by the hitchhiker? Considerations such as these must be thoroughly examined before implementation of a hitchhiking program. Ad hoc pooling in the form of regulated hitchhiking may prove to be a viable transportation system, especially in communities with a large population of non-automobile-driving people, such as a college community. However, the effects of such a program on other transportation systems, especially bus systems, need further research. Moreover, because the hitchhiker typically does not own an automobile, the energy savings that could be realized from regulated hitchhiking programs is questionable. Thus truly to conserve energy, a new breed of hitchhiker—the automobile driver—would have to be attracted. Incentives beyond that of a free ride would be necessary. Further, the probability of obtaining a ride would have to be great; automobile drivers could not be expected to forego their reliable automobiles for the mere possibility of a free ride.

Increasing the Use of Other Modes of Transportation

In addition to the use of priority lanes for buses discussed earlier (cf. Rose & Hinds, 1976), specific reinforcement contingencies have been applied to increase bus ridership, most notably by Everett and his colleagues (Deslauriers & Everett, 1977; Everett, 1973; Everett, Deslauriers, Newsom, & Anderson, 1978; Everett,

Hayward, & Meyers, 1974). In the first attempt, Everett (1973) employed both prompting and positive reinforcement procedures to encourage bus ridership. More specifically, one of the eight campus buses at Penn State University was selected as an Experimental Bus. Instructions identifying a special contingency on the Experimental Bus were advertised in the university daily paper. The contingency was that one would receive money upon boarding the "Red Star Bus"—so named because of the three red stars placed on the Experimental Bus. Quarters were used for reinforcers and were delivered by the experimenter, who stood directly behind the bus driver's seat. An ABACA design was used. Throughout the experiment, riders paid the regular 10¢ fare. For 20 consecutive weekdays, counts of baseline ridership were made. Then red stars were placed on the bus, and after passing through the turnstile, individuals were handed 25¢ and were thanked for riding the bus. This intervention lasted for 3 days before the red stars were removed and the contingency withdrawn. On the 7th day of this second baseline, the advertisement was placed in the university newspaper; the next day, the red stars were again placed on the bus, and the 25¢ reinforcement was reinstituted for 1 day, followed by a 3-day follow-up during which the red stars were removed and no advertisement was run in the paper nor reinforcement delivered. Results showed that on the 3rd day of initial intervention, a 149% increase over mean initial baseline ridership was obtained. With the 25¢ and prompt intervention, a 213% increase over initial baseline and a 180% increase over second baseline were obtained.

A number of criticisms may be leveled at the Everett study. First, the amount of money that would be received was apparently not specified in the newspaper advertisements. Many riders may have taken the Red Star Bus in hopes of receiving a sum in excess of 25¢. Second, the two treatment and follow-up conditions were very short. Indeed, the duration of both treatment phases may have been constrained by the prohibitive cost of reinforcement (25¢ paid to 370 riders during the single day of reinforcement). Further, the relatively short follow-up periods did not result in a stable rate of ridership. Third, no records were made of ridership on the other seven buses, so it is not known if the contingencies attracted individuals who normally drove a car, walked, or rode another bus. Finally, it is obvious that such a procedure is not very cost-effective. Paying people to ride a bus cannot be continued for long periods of time.

Deslauriers and Everett (1977) introduced a token reinforcement procedure in order not only to increase bus ridership but to avoid the prohibitive costs of 25¢ rewards. In this study, a control bus was used that traveled the same route as the Experimental Bus. As in the previous study, the Experimental Bus was identified with three red stars, and during treatment an advertisement that described the Experimental Bus route and the redeemable tokens was placed in the daily campus newspaper. A handout that explained the experimental contingencies and the goods and services for which the tokens could be exchanged was made available on the Red Star Bus. Several businesses participated in the study by offering to exchange tokens for food, movies, records, and so forth. At the end of the study, these businesses were reimbursed according to a previously established rate of exchange per token, which varied from 5¢ to 10¢. Questionnaires administered periodically throughout the study assessed the origin, destination, length, and purpose of each ride in addition to demographic characteristics and mode of transportation exchanged for bus riding.

Following 6 weeks of baseline, a VR-3 token reinforcement schedule was introduced for 15 days on the Experimental Bus. During this time, large wooden stars were mounted on the bus as well as a sign that displayed information on the probability of receiving a token. This information also appeared thrice weekly in the daily campus newspaper. Following the VR-3 condition, CRF was introduced for 10 days. Then a VR-3 schedule identical to the first VR-3 condition was reintroduced for 8 days, followed by 10 days of follow-up during which all contingencies, red stars, and advertisements were withdrawn. The first VR-3 schedule provided for an evaluation of intermittent token reinforcement independent of continuous token reinforcement.

Ridership was analyzed for the last 6 days of each condition. A 30% increase over initial baseline was obtained during the first VR-3 condition. This increase was maintained during the CRF and during the second VR-3 condition. During follow-up, bus ridership declined toward levels on the control bus. The vast majority of bus riders reported on the questionnaires that they would have walked had they not taken the bus. Thus although token reinforcement did increase bus ridership, it failed to attract drivers away from their cars. Deslauriers and Everett noted that their campus setting is really inadequate for evaluating the impact of token reinforcement on car drivers because of university restrictions on traffic. Unfortunately, the nature of these restrictions was not specified by the authors. If it is merely a prohibition against freshmen owning cars, as is typical of many universities, then it is not unreasonable to ask why the contingencies did not attract upperclassmen, graduate students, faculty, and staff who drive automobiles.

A major drawback of the Deslauriers and Everett study is that the authors failed to report the cost of reimbursement to merchants who participated in the token exchange so that a direct cost comparison could be made between the CRF and VR-3 schedules. Instead, Deslauriers and Everett reported that if only fare revenue and direct cost of reimbursement are considered, then the VR-3 conditions resulted in a net loss in revenue of 5%, whereas the CRF resulted in a 75% net loss in revenue. The practice of reporting costs in different terms from study to study makes direct comparison of intervention costs per unit of behavior a cumbersome and time-consuming task, especially for the nonmathematically inclined. Here, as in the Hake and Foxx (1978) study, is a demonstration of the need for a standardized format to report cost-effectiveness.

The behavioral community research by Everett and his students demonstrates clearly that bus ridership can be increased through the application of token reinforcement. However, the interventions failed to attract drivers away from their cars and failed to be cost-effective. The failure of this approach to reduce automobile usage may rest solely on the uniqueness of the campus environment and the student population. Similar studies using central business districts and suburban settings are obviously needed. If such studies also fail to attract the automobile driver, then it may be that the Everett and Hayward (1974) "Cell 3 approach" (of increasing the positive consequences for not-car-driving) is just not cost-effective in inducing people to change their mode of transportation from the convenient automobile to some energy-conserving alternative. Perhaps the secondary reinforcers for which bus tokens could be exchanged need to be altered in order to encourage a change in transportation mode. For example, a weekly lottery with a large sum as the prize (e.g., $100) may be a cost-effective, positive consequence of not-car-driving.

SUMMARY AND SUGGESTIONS

Table 3.4 summarizes the research reviewed in this chapter in terms of the targets and outcomes of both antecedent and consequent strategies for encouraging energy-conserving transportation. The research reviewed herein demonstrates the potential for the application of behavioral technology to reduce automobile usage and to increase the use of other more energy-conserving modes of transportation. The identification of transportation contingencies and their experimental validation is still in the initial stages. The influence of direct, personal cost on transportation behavior and mode selection is not clear. Although an awareness of daily operating cost may increase negative attitudes toward the automobile as a viable mode of transportation, it apparently does not reduce mandatory automobile usage (Hayward & Everett, 1976). The findings of Golob et al. (1972) suggest that for at least one mode of transportation, immediate daily fare collection is most preferred. The influence of direct cost on nonmandatory trips has not been investigated.

Reduced VMT has been achieved through monetary rewards (Foxx & Hake, 1977; Hake & Foxx, 1978) and public feedback and team competition (Reichel & Geller, 1980). Increased mpg has been achieved through feedback (Lauridsen, 1977; Runnion et al., 1978) and social recognition and lotteries (Runnion et al., 1978).

Increases in carpooling have been obtained by environmental interventions such as priority lanes (MacCalden & Davis, 1972; Rose & Hinds, 1976), preferential parking (Hirst, 1976; Pratsch, 1975), monetary rewards (Letzkus & Scharfe, 1975), organizational matching efforts (Jones & Derby, 1976), and a combination of preferential parking and token rewards (Jacobs et al., 1979). Significantly influencing carpooling are such demographic factors as age and reference-group affiliation (Margolin et al., 1976), as well as such psychosocial factors as perceived inconvenience, loss of independence, social skills (Barkow,

TABLE 3.4
Transportation Management for Energy Conservation

	Antecedents			Consequences		
Behavioral Community Research	*Target*	*Strategy*	*Outcome*	*Target*	*Strategy*	*Outcome*
Everett (1973) (college campus)				bus ridership	monetary reward	149% increase
	bus ridership	newspaper ad	combined with consequences (cwc)	bus ridership	monetary reward	213% increase
Everett et al. (1974) (college campus)	bus ridership	newspaper ad	cwc	bus ridership	token economy (CRF schedule)	50% increase[a]
Deslauriers & Everett (1977) (college campus)	bus ridership	newspaper ad	cwc	bus ridership	token economy (VR 3 sched.) (CRF sched.) (VR 3 sched.)	+30%[a] +30% +30%
Rose & Hinds (1976) (interstate highway)	bus ridership	lane-designation sign	cwc	bus ridership	priority lane[b]	500% increase
	carpooling			carpooling	priority lane[c]	+11% in persons/ vehicle
Letzkus & Scharfe (1975) (organization with 685 employees)	?	?	?	carpooling	reduced parking fee 4+ = free parking 3 = 1/3 fee 2 = 2/3 fee	participation rates were: 23.1% 6.6% 3.5%
Pratsch (1975) (various organizations)	?	?	?	carpooling	preferential parking	44% increase

(Continued)

TABLE 3.4
(Continued)

Behavioral Community Research	Antecedents			Consequences		
	Target	*Strategy*	*Outcome*	*Target*	*Strategy*	*Outcome*
Hirst (1976) (Washington, D.C., parking lots)	?	?	?	carpooling	preferential parking	22.1% increase
					preferential parking & $2 daily cost to solo draivers	43.8% increase
Jacobs et al. (1979) (college campus)	carpooling	flyer, computer-match list, program prompt	cwc	carpooling	token reinforcement & reserved parking	+8% (lot A) + 3% (lot B)
					reserved parking	+10% (lot A)
Foxx & Hake (1977) (college campus)	reduced VMT	goal setting (pamphlet)	cwc	reduced VMT	monetary reward, car servicing tours[d]	20% reduction
Hake & Foxx (1978) (college campus)	reduced VMT	goal setting (pamphlet)	cwc	reduced VMT	monetary reward[d]	20% reduction
					self-monitoring and monetary reward[e]	12.6% reduction
Reichel & Geller (1980) (college campus)	reduced VMT	verbal instructions, flyer	cwc	reduced VMT	group competition[d] & public feedback	22% reduction/ 62% reduction

(Continued)

TABLE 3.4
(Continued)

Behavioral Community Research	*Antecedents*			*Consequences*		
	Target	*Strategy*	*Outcome*	*Tartet*	*Strategy*	*Outcome*
Lauridsen (1977) (organization)				increased mpg	mpg feedback meter	+.65 mpg
					mpg feedback meter, lottery	+.66 mpg
Runnion et al. (1978) (organization)	increased mpg	written instructions	cwc	increased mpg (1st year)	public feedback, personal recognition, social reinforcement, peer competition	+5.06%[f] (short haul) +2.08% (long haul)
	increased mpg	written instructions	cwc	increased mpg (2nd year)	public & personal feedback, social reinforcement, lotteries	+4.18%[f] (short haul) +8.96%[f] (long haul)

[a] Questionnaire results revealed that most new bus riders would have walked; car drivers were not modified.
[b] Accompanied by 740% increase in number of bus trips during rush hour.
[c] Enforcement problems–six police officers for 5.5 miles.
[d] Confounded by "leader variable"–subjects associated with investigator.
[e] Monetary reward for calculating average daily VMT.
[f] Only 1 week of initial baseline.

1974; Horowitz & Sheth, 1977), sex of riders, and solidarity among riders (Dueker & Levin, 1976).

Vanpooling is a particularly cost-effective application of contingency management that has reduced automobile usage over extended periods of time for a variety of populations and settings (e.g., Forstater & Twomey, 1976; Owens & Sever, 1977). Regulated hitchhiking (Barkow, 1974; Dallmeyer, 1975) may provide transportation for people in need of a ride, but the potential energy savings of such a program would require a considerable effort to attract automobile drivers to hitchhike.

Although bus ridership has been increased with a variable-ratio, token reinforcement schedule (Deslauriers & Everett, 1977), the program was not very cost-effective, and automobile drivers were not attracted to the bus. Actually, priority lanes for buses and increased bus service have shown greater success in attracting automobile drivers (Rose & Hinds, 1976) than have the procedures applied by Everett et al.

Presuming a goal to reduce automobile usage, the present review indicated the following suggestions for future research.

1. Research should be directed at noncollege populations in noncollege environments. The uniqueness of the university population and setting sets the occa sion for transportation needs that differ markedly from those of the worker and household manager. Additionally, reduction of nonessential driving should be stressed when viable alternative modes are lacking for essential trips.

2. A community transportation package should be considered, combining incentives both for reduced automobile usage and for use of alternative modes of transportation. Such a comprehensive approach could offer small incentives for automobile mileage reduction and small incentives for the use of alternate modes, with maximum incentives for reduced automobile usage *and* use of alternate modes. As Margolin et al. (1976) suggested, a comprehensive approach should consider the various populations involved and derive incentives with those subgroups in mind.

3. Because of the difficulty in achieving cost-effective interventions, the effects of nonmonetary contingencies such as social recognition and the use of leader models and prompts should be investigated. Further, the development of new, energy-conserving social norms should be stressed.

4. The application of lotteries and contests may provide cost-effective strategies for motivating energy-conserving transportation.

5. Employer-sponsored carpool and vanpool programs have tremendous potential for energy savings and have been successful in attracting solo drivers to ride sharing. The important question of organizational and matching efforts versus incentives-only programs must be answered. A great deal of time and effort may be wasted if incentives alone can be as effective as more cumbersome matching efforts. Further, employers have the unique capability of designing incentive programs around employee characteristics.

6. Less formalized alternatives to solo driving, such as ad hoc carpooling, though capable of meeting transportation needs, may not have a substantial impact on energy savings. Indeed, the mass-transit system of a community may suffer from such programs.

ONE FINAL CONSERVATION STRATEGY

An alternative approach to the general target of reduced automobile usage is to decrease the need for mobility by making the home environment more reinforcing and efficient in meeting family needs (cf. Skinner, 1975). In an attempt to modify travel behavior, Gurin (1976) investigated several strategies to effect changes in the home environment. The eight strategies investigated were: (1) substitute phone calls for trips; (2) reduce household and personal stress; (3) improve dwelling designs to lessen family conflicts, thereby reducing the need for trips to "get away from it all"; (4) increase home-centered recreational activities; (5) facilitate home delivery of goods and services; (6) locate desired destination activities nearer to home; (7) improve household location decisions; and (8) reduce peak-hour travel.

Social science literature and information from a comprehensive case study of male high-school students from the Boston working class were used to forecast behavioral responses to these eight strategies. Gurin (1976) found substantial institutional and cultural barriers to the implementation of these strategies as well as psychosocial barriers. Much like Margolin et al. (1976), Gurin demonstrated the need to consider subgroup characteristics before implementing specific interventions. For example, Gurin considered substituting telephone calls for travel. He noted the following barriers to such an intervention applied especially to working-class teenagers: (1) parental reluctance to install telephones in rooms where teenagers could hold private conversations; (2) teenagers value face-to-face interactions in order to develop and strengthen friendships; (3) teenager interactions are felt to be more effective in person; (4) teenage social conversations often take place in groups; and (5) teenage panelists expressed distrust of the telephone company. None of these barriers specifically preclude the application of contingency management to teenage populations in order to encourage telephone calls in place of automobile usage. Rather, these barriers present specific factors that incentives should be designed to counter.

CONCLUSION

Attempts to reduce energy consumption through operant technology have demonstrated that transportation behaviors can be modified toward energy conservation. However, many of the interventions were not cost-effective. This raises the critical issue of cost-feasibility for long-term, large-scale applications of individual incentives. Individual incentives, ranging from monetary rewards to reduced work-trip expenditures and to tokens that can be exchanged for goods and services, are just not practical for long-term behavior change. Energy-conserving behaviors are lifelong behaviors that must be maintained, even when there are no immediate energy shortages, lest shortages develop. Ideally, social norms need to be modified so that people will want to carpool and use mass transit (Hirst, 1976; Stern, 1979; Stern & Kirkpatrick, 1977). In other words, to create a truly energy-conserving community, conserving behaviors should be expected, socially valued, and socially rewarded. Such norms may even be internalized by individuals so that conserving behaviors become intrinsically reinforced. One approach toward this goal may be to stress group incentives rather than individual incentives (e.g., Reichel & Geller, 1980). Labor unions, work groups, institutions, community clubs, and neighborhoods could become targets for behavioral interventions.

Consequently, applied behavioral analysis is a promising approach for the reduction of energy consumption through the modification of transportation-related behaviors. To achieve energy savings, there are specific needs within the community for the skills of the behavioral scientist as well as for the transportation engineer. It is imperative that both the behavioral scientist and the engineer acquaint themselves with the skills, knowledge, capability, and technology of the other in order to implement and evaluate cost-effective, communitywide transportation programs for reducing energy consumption.

ACKNOWLEDGMENTS

Preparation of this chapter was supported by two Title I, HEA Grants for Community Service and Continuing Education in Virginia: (1) Community Education Model for Energy Conservation, and (2) Community Action Model for Energy Conservation. Although the projects represent a collaborative, interdisciplinary effort of faculty and students from Architecture, Civil Engineering, Education, Physics, and Psychology, only the authors of this chapter should be held responsible for its content.

REFERENCES

American Automobile Association. *Gas watchers' guide*. Falls Church, Va.: Author, 1975.

Andrle, S., & Dueker, K. J. *Attitude toward and evaluation of carpooling* (Tech. Rep. No. 32). Iowa City: Center for Urban Transportation Studies, University of Iowa, 1974.

Applied Decision Systems Inc. *Transportation planning simulation*. Report presented at Environmental and Urban Systems Research Conference, U.S. Department of Transportation, Washington, D.C., September 21, 1970.

Barkow, B. *The psychology of car pooling*. Ontario, Canada: Ministry of Transportation and Communication, 1974.

Barron, Lieutenant. Department of Police, City of Eugene, Oregon. Personal communication, August 16, 1978.

Bent, M. D. *Vanpooling: A technique for estimating impacts* (Tech. Memo. 34). Baltimore, Md.: Regional Planning Council, June 1977.

Brasted, B. Personal communication, July 25, 1978.

Bryant, D. A. *Study and evaluation of commuter carpool programs in certain metropolitan areas* (GCA-TR-74-8-G). Bedford, Mass.: GCA Corporation, Technology Division, April 1974.

Carter, President Jimmy. *The national energy plan.* April 29, 1977.

Dallmeyer, K. E. *Hitchhiking—A viable addition to a multimodal transportation system.* Boulder: Center for Transportation Studies, University of Colorado, March 1975. (NTIS No. PB-242 203)

Deslauriers, B. C., & Everett, P. B. Effects of intermittent and continuous token reinforcement on bus ridership. *Journal of Applied Psychology,* 1977, *62,* 369–375.

Dorf, R. C. *Technology and society.* San Francisco: Boyd and Fraser, 1974.

Dueker, K. J., & Levin, I. P. *Carpooling: Attitudes and participation* (Tech. Rep. No. 81). Iowa City: Center for Urban Transportation Studies, University of Iowa, 1976.

Environmetrics, Inc. *The urban systems model.* Washington, D.C.: Office of The Secretary, Assistant Secretary for Environment and Urban Systems, U.S. Department of Transportation, 1972.

Everett, P. B. The use of the reinforcement procedure to increase bus ridership. *Proceedings of the 81st Annual Convention of the American Psychological Association,* 1973, *8*(2), 891–892.

Everett, P. B., Deslauriers, B. C., Newsom, T., & Anderson, V. B. The differential effect of two free ride dissemination procedures on bus ridership. *Transportation Research,* 1978, *12,* 1–6.

Everett, P. B., & Hayward, S. C. Behavioral technology—An essential design component of transportation systems. *High Speed Ground Transportation Journal,* 1974, *8,* 139–143.

Everett, P. B., Hayward, S. C., & Meyers, A. W. Effects of a token reinforcement procedure on bus ridership. *Journal of Applied Behavioral Analysis,* 1974, *7,* 1–9.

Everett, P. B., Studer, R. B., & Douglas, T. J. Gaming simulation to pretest operant-based community interventions: An urban transportation example. *American Journal of Community Psychology,* in press.

Federal Energy Administration. *Tips for energy savers.* Washington, D.C.: U.S. Government Printing Office, 1977.

Ferster, C. B., & Skinner, B. F. *Schedules of reinforcement.* New York: Appleton-Century-Crofts, 1957.

Fleischman, E. Public incentives. *Proceedings of the 1975 National Conference on Areawide Carpooling,* Houston, Texas, December 8–10, 1975. Washington, D.C.: Federal Highway Administration, Urban Planning Division, 1975.

Forstater, I., & Twomey, E. *Vanpooling: A summary and description of existing vanpool programs.* Washington, D.C.: U.S. Environmental Protection Agency, January 1976.

Foxx, R. M., & Hake, D. F. Gasoline conservation: A procedure for measuring and reducing the driving of college students. *Journal of Applied Behavior Analysis,* 1977, *10,* 61–74.

Geller, E. S., Chaffee, J. L., & Ingram, R. E. Promoting paper-recycling on a university campus. *Journal of Environmental Systems,* 1975, *5,* 39–57.

Golob, T. F., Canty, E. T., Gustafson, R. L., & Vitt, J. E. An analysis of consumer preferences for a public transportation system. *Transportation Research,* 1972, *6,* 81–102.

Gurin, D. B. Pragmatic evaluation of telephones, activity scheduling, and other strategies to modify travel behavior of population subgroups. *Transportation Research Record,* 1976, No. 583, 29–35.

Hake, D. F., & Foxx, R. M. Promoting gasoline conservation: The effects of reinforcement schedules, a leader and self-recording. *Behavior Modification,* 1978, *2,* 339–369.

Hargroves, B. T., & McGray, M. *TRANSPON (transportation policy negotiation).* Unpublished paper, The Pennsylvania Transportation Institute, The Penn State University, University Park, Pennsylvania, 1973.

Hayward, S. C., & Everett, P. B. *A failure of response cost feedback to modify car driving behavior.* Paper presented at the Annual Meeting of the Midwestern Association of Behavioral Analysis, Chicago, Illinois, May 1–4, 1976.

Hirst, E. Transportation energy conservation policies. *Science,* 1976, *192*(4234), 15-20.

Horowitz, A. D., & Sheth, J. N. *Ridesharing to work: A psychosocial analysis.* Warren, Mich.: General Motors Corporation, 1977.

Hurley, J. Personal communication, July 11, 1978.

Ingram, R. E., & Geller, E. S. A community-integrated, behavioral modification approach to facilitating paper recycling. *JSAS Catalog of Selected Documents in Psychology,* 1975, *5,* 327. (Ms. No. 1097)

Jacobs, H., Fairbanks, D., Poche, C., & Bailey, J. S. *Behavioral community psychology: Multiple incentives in encouraging carpool formation on a university campus.* Unpublished manuscript, Florida State University, 1979.

Jones, B., & Derby, J. Sacramento car-pool project: Interim evaluation report. *Transportation Research Record,* 1976, No. 619, 38-42.

Lauridsen, P. K. *Decreasing gasoline consumption in fleet-owned automobiles through feedback and feedback-plus-lottery.* Unpublished thesis, Drake University, May 1977.

Letzkus, T., & Scharfe, V. Employer incentive programs. *Proceedings of the 1975 National Conference on Areawide Carpooling,* Houston, Texas, December 8-10, 1975. Washington, D.C.: Federal Highway Administration, Urban Planning Division, 1975.

MacCalden, M., & Davis, C. *Report on priority lane experiment on the San Francisco-Oakland Bay bridge.* San Francisco, Calif.: Department of Public Works, 1972.

Margolin, J. B., Misch, M. R., & Dobson, R. D. Incentives and disincentives to ride-sharing behaviors: A program report. *Transportation Research Record,* 1976, No. 592, 41-44.

Miller, C., & Deuser, R. Issues in enforcement on busway and bus and car-pool lane restrictions. *Transportation Research Record,* 1976, No. 606, 12-17.

Nietzel, M. T., Winett, R. A., MacDonald, M. L., & Davidson, W. S. *Behavioral approaches to community psychology.* New York: Pergamon Press, 1977.

Owens, R. D., & Sever, H. L. *The 3M commute-a-van program* (Progress Report II). St. Paul, Minn.: 3M Company, 1977.

Pratsch, L. *Carpool and buspool matching guide* (4th ed.). Washington, D.C.: U.S. Department of Transportation, Federal Highway Administration, January 1975.

Pratsch, L. *Vanpooling discussion paper.* Washington, D.C.: Federal Highway Administration, Unpublished paper, April 1977.

Reichel, D. A., & Geller, E. S. Group vs. individual contingencies to conserve transportation energy. Paper presented at the *26th Annual Meeting of the Southeastern Psychological Association,* Washington, D.C., March 26-29, 1980.

Reichel, D. A., & Geller, E. S. *Behavioral approaches to conserve transportation energy.* Proceedings of the Annual Southeastern Regional Meeting of the Society for General Systems Research, Blacksburg, Va., April 18-20, 1979.

Rice, B. Fighting inflation with buttons and slogans. *Psychology Today,* 1975, *7,* 49-52.

Rose, H. S., & Hinds, D. H. South Dixie Highway contraflow bus and car-pool lane demonstration project. *Transportation Research Record,* 1976, No. 606, 18-22.

Runnion, A., Watson, J. D., & McWhorter, J. Energy savings in interstate transportation through feedback and reinforcement. *Journal of Organizational Behavior Management,* 1978, *1,* 180-191.

Scheiner, J. L., & Keiper, S. A. Car-pool information project: Innovative approaches improve results. *Transportation Research Record,* 1976, No. 619, 16-18.

Skinner, B. F. *Walden two revisited* (invited address). 83rd Annual Convention of the American Psychological Association, Chicago, Illinois, September 1975.

Stern, P. C. The car-pool game: How to mobilize group spirit. *Psychology Today,* 1979, *12,* 16; 26.

Stern, P. C., & Kirkpatrick, E. M. Energy behavior. *Environment,* 1977, *19,* 10-15.

United States Department of Transportation. *Carpool incentives and opportunities.* Washington, D.C.: U.S. Government Printing Office, February 1975.

Whitlow, D., & Manzanares, D. Memorandum to Bill Lamont, acting City Manager of Boulder, Colorado, June 11, 1973.

Witmer, J. F., & Geller, E. S. Facilitating paper recycling: Effects of prompts, raffles, and contests. *Journal of Applied Behavior Analysis,* 1976, *9,* 315–322.

Yukobousky, R., & Fichter, P. Mobility club: A grassroots smalltown transport concept. *Transportation Research Record,* 1976, No. 559, 89–100.

4 Encouraging Residential Energy Conservation Through Feedback[1]

Clive Seligman
University of Western Ontario
Lawrence J. Becker
City University of New York
John M. Darley
Princeton University

INTRODUCTION

The seriousness of the energy problem has been dramatically characterized by President Carter, who said: "With the exception of preventing war, this is the greatest challenge our nation will face during our lifetime." The present chapter discusses some of the possible points of application of social science to the problem. We begin by sketching for the reader the pattern of energy usage in the United States today and the prevailing estimates of how much of that energy might be conserved. Both the level of consumption and estimates of conservation are startlingly high: The United States consumes what is increasingly perceived as an inordinate amount of the world's energy supply, and it is argued that a great deal of that energy can be saved. Next, after residential energy consumption has been fitted into the national energy-use profile, we analyze what levels of savings are possible from various alterations in homeowners' behaviors. We then report some of our research that uses psychological techniques—primarily timely and interpretable consumption feedback—to reduce homeowners' energy consumption. Finally, we return to the more general level and suggest other points of application of social science to the energy conservation problem.

[1]Correspondence regarding this chapter should be sent to Clive Seligman, Department of Psychology, University of Western Ontario, London, Ontario, Canada. N6A 5C2.

The impact of energy on our society is pervasive. It is not simply a narrow problem concerning whether there will be enough natural gas available to heat the homes of the Northeast—as was the situation during the 1976–1977 winter—but a fundamental problem affecting the economy; balance of payments; foreign relations; despoliation of land, air, and water; and, in the case of nuclear power plants, even our safety. One indication of the importance of energy is the close relationship between gross national product and energy consumption. This strong relationship is evident both within the United States as it industrialized over time (see Fig. 4.1) and across the less industrialized countries at one point in time (see Fig. 4.2).

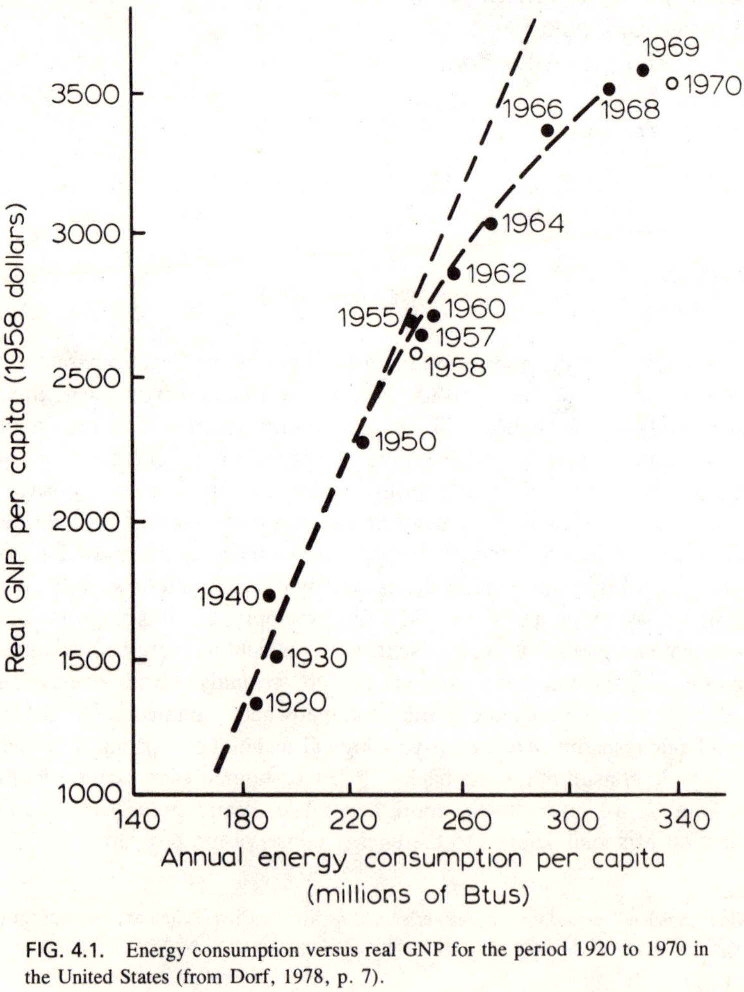

FIG. 4.1. Energy consumption versus real GNP for the period 1920 to 1970 in the United States (from Dorf, 1978, p. 7).

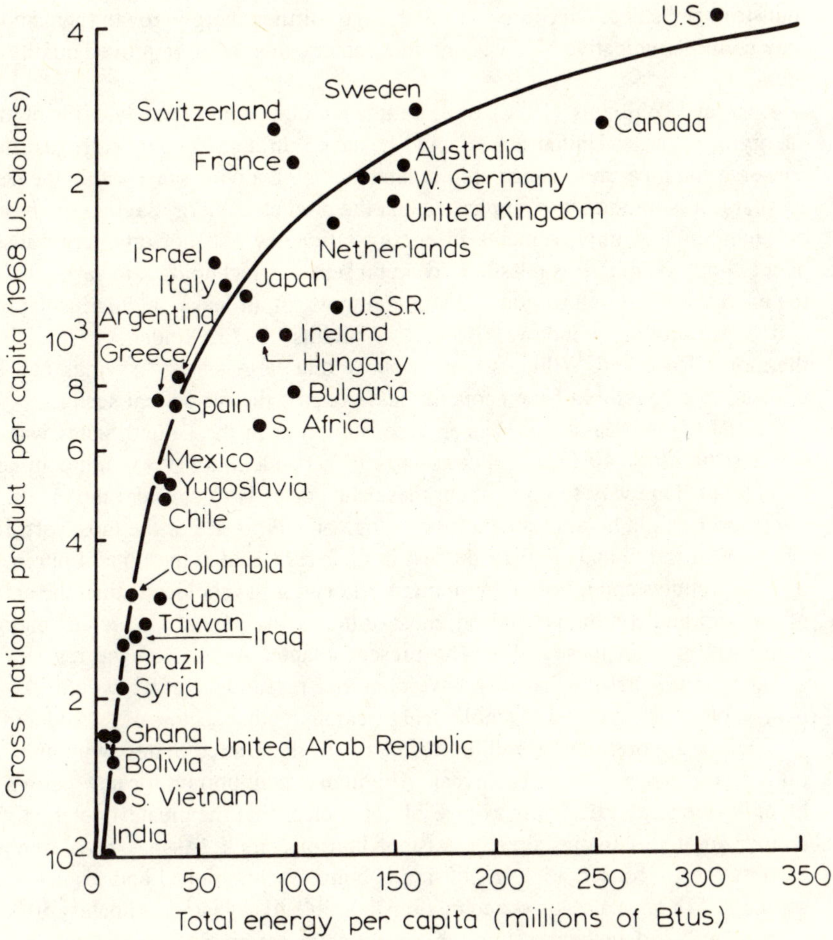

FIG. 4.2. Energy use per capita versus GNP per capita in 1968 for several nations (from Dorf, 1978, p. 5).

As Fig. 4.2 also shows, energy is consumed unequally in the world. The United States consumes about a third of the world's energy with only 6% of its population. More distressingly, among the most highly industrialized countries, it is not clear that a nation's well-being (as measured by GNP) is significantly enhanced by greater energy consumption. For example, one can see from Fig. 4.2 that Canada and Switzerland have similar per capita GNPs. Yet Canada expends more than twice as much energy per capita as does Switzerland. Only the most ethnocentric Canadian would argue that the quality of life in Canada is more than twice as good as it is in Switzerland. Thus, although increased energy consumption is clearly crucial to a nation's economy as it grows to an advanced industrialized stage, once it reaches that stage, further energy growth (per capita) may be more indicative of waste and inefficiency than of an improved quality of life.

Ross and Williams (1976) have estimated that approximately 50% of the energy used in the United States could be saved through economically justified conservation measures, which are implementable on a wide scale within the next 15 years. It is important to emphasize that the predicted energy savings are based on technological improvements in energy efficiency and not on energy abstinence. For example, it is possible, using present-day technology, to save 50% of the energy needed to heat houses by plugging warm-air leaks, adding insulation, and by improving windows (Ross & Williams, 1976; Sinden, 1978). Furthermore, Ross and Williams (1976) show that large energy savings can be realized in all sectors of the economy and not just the residential sector.

In 1973, the breakdown of energy consumption in the United States was as follows: industry (40%), transportation (25%), residential (19%), and commercial (16%). One way to think about these categories is to consider the extent to which each might be amenable to psychological solutions. For the most part, the behavior of individuals in their homes, in their cars, and at the time of purchase of new products appears more tempting territory for psychologists than the arena of industry and commerce. Indeed, most of the psychological research on energy conservation is in these areas. The present chapter focuses on the residential sector because that is where we have done our research. But we would like to make a plea for increased psychological research in other sectors of the economy.

There are more than 60 million dwellings in the United States, with about 2 million new housing starts each year. Although it is important for new houses to be built as energy-efficiently as possible, it is clear that the greatest potential for conservation lies in the already existent housing stock. As has been implied earlier, energy conservation in housing is both a technological and a behavioral problem. We have already stated that it is possible to save approximately 50% of the energy used in houses through appropriate engineering practices. Can as much energy be saved by bringing psychological insight to bear on this problem? We would like to suggest that this question is put misleadingly. If a person has been persuaded to buy and install attic insulation that results in an energy saving, who should take credit—the psychologist-persuader or the manufacturer of insulation?

A distinction can be made between attempts to influence people to make the necessary one-shot changes to their homes—for example, adding insulation—and attempts to persuade people to behave differently in their homes—for example, lowering thermostats. The research reported here concerns continual behavioral changes, as does much of the research in the area (cf. Winett & Neale, 1979).

In exhorting homeowners to adopt an energy-conserving style of life, it is important to realize that not all energy-conserving behaviors lead to the same savings. The big users of energy in a house are the furnace, air conditioner, water heater, and refrigerator. In a typical house in New Jersey, residents would have to turn off all their lights for about a day to save as much energy as they would by using their air conditioner 1 hour less. Clearly, our behavioral strategies must be targeted to the right behavior. The payoff is much bigger if we concentrate on air-conditioning than on lighting. In the homes we have studied in New Jersey, a 1-degree change in the thermostat setting for central air-conditioning would result in a 12% saving in air-conditioning use and a 7% to 8% saving in total electricity use. This would translate into savings of about 325 kwh of electricity per house over the summer months. If these savings could be achieved in only 1 million of the 60 million dwellings, enough energy would be saved to supply all the electricity required to run more than 33,000 additional homes for 1 year. Thus, a very modest change in thermostat setting, which would have negligible effects on personal comfort, could result in dramatic energy savings, even if less than 2% of the population made the change.

One might think that as houses are better built, the role of human behavior would become less important. However, data collected by the Twin Rivers project (Socolow, 1978) show this not to be the case. In houses that have been successfully made more energy-efficient (with 20% to 25% savings), the variance in energy consumption among the houses remained almost the same as it was before the retrofits took place (Woteki, 1977). Other findings from the Twin Rivers project emphasizing the importance of resident behavior show that winter gas consumption varies as much as two to one and that summer electricity consumption varies as much as three to one in samples of identical townhouses. As these houses are identical in floor plan, position in the interior of a townhouse row, builder, construction materials, and climate; and since there is a good deal of homogeneity among the families on a number of major demographic variables, it is likely that most of the variation in consumption is due to the behavior of the people in the houses. Also, where there has been a change in residents, it has been found that the energy consumption of the house with the new residents living in it cannot be predicted from the consumption of the same house with the previous residents (Sonderegger, 1978).

CONSUMPTION FEEDBACK

Given these findings about the magnitude of residential energy consumption and the patterning of its usage, we decided to begin our own research with an attempt to reduce residential energy consumption. In thinking about psychological procedures to modify residential energy consumption, we became aware of several issues: First, in residential energy use, we have to deal with a whole family and not with just one individual. Thus, procedures need to be developed that influence all family members. Simply showing up at the front door and delivering a persuasive message to the person answering the door will not work. Second, we are concerned with behavior that occurs over time. Consuming energy is a continual process. Ways need to be found to maintain people's interest in saving energy over time. Third, consuming energy is not a highly salient activity. As the family goes about its business of getting the kids off to school, cooking, and so on, it is consuming energy without much thought. How can energy consumption in the home be made more salient without having a crisis atmosphere? Fourth, although residents must be made more aware of their energy consumption, any procedure for helping them save energy must not be too intrusive or disruptive of their normal everyday routine. A conservation strategy that causes too much inconvenience or discomfort will not be accepted. Fifth, despite all the protests about the cost of energy, energy is still a very good buy. It amounts only to about 5% of yearly family income, although this percentage will rise in the coming years. Any procedure for facilitating conservation must itself not involve too great an expense. In sum, we need to intervene in a way that involves the whole family, maintains interest in conservation over time, makes energy use and its potential savings salient, and is not too inconvenient or expensive.

One procedure that meets the conditions outlined is feedback—giving people frequent information about their rate of energy use. Consider a typical house: Appliances are run, hot water is used, the air conditioner cycles, and the homeowner has no idea how much energy he or she is using. The utility bill comes, at best, once a month and often is an estimate rather than being based on actual use. Clearly the homeowner lives in an information-deficient energy environment. What would happen if the homeowner were given more information about his or her energy consumption, if the feedback loop between the homeowner and the house were closed?

Interest in feedback as a possible conservation aid was stimulated by the large body of psychological research dating as far back as the early 1900s and showing that when people are given feedback, their performance generally improves. This has been found to be the case for a wide variety of tasks, ranging from solving simple addition problems to loading logging trucks, from drawing lines of a given length to tracking and ranging tasks involving anti-aircraft guns. (See Ammons, 1956, and Bilodeau & Bilodeau, 1961, for reviews of this literature.) Given these findings, it seemed reasonable to ask whether feedback would also facilitate performance when the task involved energy conservation.

Feedback, rather than some other promising energy conservation technique, was chosen as the subject of our first experiments because of our belief that a successful feedback procedure would play a considerable role in any conservation plan. There may be many different diets for losing weight, but they all require a bathroom scale to be used frequently to monitor the progress of the diet. We suspected that not only would feedback by itself be a successful conservation strategy but also that it would be a necessary tool to convince homeowners that a recommended conservation program could work.

A good example of how frequent feedback can affect energy consumption is found in a study by Seligman and Darley (1977). Their study was an attempt to reduce electricity consumption. Electricity consumption was examined because the study was conducted during the summer when, in the studied houses, 70% of the electricity used is for central air-conditioning. Moreover, air-conditioning use could be modified to a large extent by the homeowner. The non-air-conditioning energy use is relatively constant on a daily basis. The subjects were 29 homeowners living in identical three-bedroom townhouses who volunteered to participate in an energy conservation study. The subjects, in this study as well as the others reported later, were largely young married couples, typically with two young children, one perhaps in grade school. The average husband was in his mid-thirties and his wife, in her early thirties. Almost all of the men and most of the women had college degrees. The median family income at the time our studies were conducted was in the $20,000–$25,000 range.

The homeowners were randomly assigned either to a feedback group or to a control group. Beginning in July 1975, the electric meters of all homes were read by a research assistant each weekday afternoon for a month. The average temperature was computed for each day from the hourly readings provided by on-site recording equipment. For each house, a regression line was plotted to predict daily electricity consumption from daily average temperature. The correlations were generally greater than .8. By inference, the electric meter reading was proportional to a reading that would have been taken from a meter on the air-conditioning system plus a constant. For each house, then, it was possible to predict its future rate of energy usage based on outdoor temperature.

In addition to reading the electric meters each weekday, beginning in August, the research assistant calculated the ratio of actual to predicted consumption for each house in the feedback group, Tuesday through Friday. (Predicted consumption was based on the immediately preceding 24-hours temperature readings inserted into the regression equation for each house.) This ratio was displayed as a percentage score by means of small plastic numbers placed in a Lucite frame (about 8 cm by 12 cm) that was attached to the outside of the kitchen window. The number was easily visible to the family from inside the house. For example, if the household's predicted performance was that it would use 100 kwh of electricity and it actually used 80 kwh, then its display for that day would read "80%." Note that this feedback reflected a household's rate of electricity consumption relative to a standard derived from its past consumption rather than some theoretical standard or one derived from other households.

The same day the feedback began, each household in the feedback condition received a letter explaining the feedback procedure—that is, how predictions of their electricity consumption were made and what the numbers in the Lucite frame meant. The letter also focused the homeowners' attention on air-conditioning: They were told that in the summer, the largest use of electricity was due to air-conditioning. The control group was sent the same letter except for the

TABLE 4.1
Average Daily Electricity Consumption (in kwh)

Descriptive Statistic	Before Feedback		During Feedback	
	Feedback Group	*Control Group*	*Feedback Group*	*Control Group*
M	68.3	69.1	48.6	54.3
SD	10.5	11.0	7.9	5.1

part dealing with the feedback procedure. Therefore, summarizing the similarities between conditions, all households—regardless of condition—had their electric meters read 5 days a week, were told they were in an energy study, that air-conditioning was the largest use of electricity, and that we hoped they would reduce their air-conditioning usage. Thus, both demand characteristics to reduce electricity consumption and information received about how to do it were the same for both groups. The feedback group differed from the control group in that it received the almost daily information about its consumption and an explanation of how that information was calculated and displayed. Feedback was given for 3 weeks.

The results (see Table 4.1) showed that before the feedback began, the feedback and control groups did not differ in electricity use. However, during the feedback period, the mean daily consumption of the feedback group was 10.5% less than that of the control group, a statistically significant difference. A number of studies, in addition to our own, have shown that feedback can result in reduced energy consumption (Hayes & Cone, 1977; Hayes & Cone, in press; Kohlenberg, Phillips, & Proctor, 1976; Palmer, Lloyd, & Lloyd, 1978; Winett, Neale, & Grier, 1979; Winett, Neale, Williams, Yokley, & Kauder, in press).

Possible Explanations of the Feedback Effect

Although feedback seems to be able to facilitate energy conservation, it is not at all obvious how it has such an effect. Feedback may work for several reasons: (1) It teaches people how to make appropriate conservation responses; (2) it rewards appropriate conservation behaviors, thereby increasing the likelihood of their being repeated; (3) it motivates people to try to save energy; (4) it provides information that an individual can use to evaluate his or her performance with regard to an implicit or an explicit goal. We discuss each of these possibilities in turn.

Human Factors Approach. One tradition, embodied in human factors research, considers the function of feedback as primarily the teaching of new, skilled responses (McCormick, 1976). In this tradition, feedback has been shown "to be the strongest, most important variable controlling performance and learning'' (Bilodeau & Bilodeau, 1961, p. 250). The type of learning that human factor psychologists have been concerned with consists of complex motor skills such as flying airplanes, shooting weapons, using tracking apparatus, and so forth. What these tasks have in common is that they are ''man–machine'' systems. These systems can be represented as loops that are closed by the presence of the human operator in them. Some information, such as flight altitude, could be presented to the human operator, and on the basis of this information, he or she would make some control system modification. In a usually relatively continuous fashion, the results of that modification would become apparent to the pilot. To use the familiar man–machine automobile system, the operator sees a corner around which he or she wants to turn, operates the steering wheel to do so, and sees through the windshield the effects of the action of changing the car's direction, modifying the controlling actions based on the new information now available. As anybody who has trained a novice to drive will recall with some terror, the discrete alternation of information processing and system control, with considerable time delays between them, is quite characteristic of untrained operators. The performance of a trained operator involves a much more continuous processing of information and a much smoother and more continuous flow of control action as well. One thrust of human performance research has been toward selecting out of the potentially available information those elements that are critical for the operator's successful control of the system and toward displaying those elements in a way that optimize the operator's control functions.

Does this description of the human factors perspective apply to the problem of feedback and energy conservation in the home? Are homeowners like automobile drivers who need to attend to a continuous flow of information to guide their control actions? Or have we given a better description of how the thermostat works than of how the homeowner behaves? Even if we do not take the human factors perspective literally, and only abstract from it the notion that feedback teaches homeowners something about their houses as energy-consuming systems, we would still quarrel with the explanatory value of the human factors approach in this particular situation. In our view, homeowners do not need to learn through trial and error which of their many possible attempts to conserve energy are successful, as displayed on some feedback chart. We suggest that it would be more efficient simply to tell homeowners that lowering the thermostat (in the winter) or adding insulation to the house will save energy. (There may be controversy over precisely how much energy will be saved through different conservation actions, but there is no doubt that energy will be saved.) The physics of the problem dictates what are successful and unsuccessful conservation attempts, and they are fast becoming common knowledge. Suppose a homeowner has been told by a credible source that he or she can save energy by lowering the thermostat setting. This homeowner will assuredly save energy so long as he or she does, in fact, lower the thermostat setting, regardless of whether further feedback is received. In contrast, we do not recommend that one verbally instruct an automobile driver on how to get around a corner and then prevent further feedback by blindfolding the driver. The point, of course, is that learning to conserve energy in the home is a very different sort of problem from learning how to drive a car, which is a classic human factors problem. The latter requires a sequence of continued alternations in information processing and system control, whereas the former only requires knowledge of the correct action and the appropriate follow-through (a response already in the homeowner's repertoire). This is not to say, however, that feedback should not facilitate energy conservation—only that it does not typically do so through the teaching of new, skilled responses. Perhaps feedback serves as a reward for conservation behavior.

Reinforcement Approach. As Bilodeau and Bilodeau (1961) pointed out, many investigators have used the terms *feedback* and *reward* interchangeably, although they refer to substantially different processes. We can make this clearer by looking at a study of residential energy conservation. Becker and Seligman (1978) did a study to test the effects of both feedback and a particular conservation device that they had developed. The purpose of the device was to provide a signal to homeowners when the outside temperature was below 68° *F* and their air conditioner was still running. Homeowners were informed that when the outside temperature was below 68° *F*, air-conditioning was no longer necessary, and the house could be cooled effectively by just opening the windows.

The device used a blue light that was displayed in the homeowner's kitchen. The blue light was connected to a thermostat situated on the outside wall of the house and to the air conditioner. The blue light would blink repeatedly when the air conditioner was on and when the outside temperature was below 68°*F*. The only way the homeowner could stop the blue light from blinking was to shut off the air conditioner. When the outside temperature was 68°*F* or higher, the blue light was off regardless of whether the air conditioner was on or off.

Forty residents were randomly assigned to one of four conditions: blue light plus feedback, feedback only, blue light only, and a control condition (no feedback and no blue light). The feedback was given three times a week for 1 month.

The results (see Table 4.2) showed that homeowners who had the blue-light device used significantly less electricity (15.7%) than homeowners who did not. However, in this experiment, homeowners who received feedback did not use significantly less electricity than those who did not receive feedback. Furthermore, the interaction was not statistically significant. For reasons we do not go into now, homeowners apparently did not believe the feedback. They claimed they saw no relationship between their conservation actions and their feedback scores. Consequently, they said they ignored the feedback.

The point we wish to make here is that if one thinks of feedback as a reward, then a question can be raised about why the residents did not stop responding to the blue light after they discounted the feedback. According to this perspective, if

TABLE 4.2
Average Daily Electricity Consumption (in kwh)
During the "Blue-Light" Period

Descriptive Statistics	*Signaling Device*		*No Signaling Device*	
	Feedback	*No Feedback*	*Feedback*	*No Feedback*
Adjusted score[a]				
M	18.30	18.24	20.61	22.76
SD	2.96	4.50	5.69	6.02

[a]Scores adjusted by removing the effect of the covariate, average daily consumption during the pretreatment period (June 1 to August 10, 1977).

the residents' response to the blue light is under the control of feedback qua reward, the response to the blue light should have extinguished, because ignored feedback could no longer be rewarding. For example, if a rat learns to make a particular response after a blue light begins blinking in order to obtain food, then one would correctly predict that the response to the blinking blue light would eventually extinguish after the reward (food) was withdrawn. In this example, the rat's response to the blue light has no value other than that it leads to food. When the food is no longer forthcoming, the response becomes obsolete.

However, in the Becker and Seligman (1978) study, the response to the blue light does have value in and of itself: It saves energy. The appropriateness of the resident's response to the blue light (namely, to turn off the air conditioner and open the windows) is unambiguous: The less time the air conditioner is on, the less electricity is being used. It makes sense to respond to the blue light even if feedback is absent (or ignored). The data from the condition in which only the blue light was present confirm this: In the blue-light and no-feedback condition, residents used the same low amount of electricity as did residents in the blue-light and feedback condition. We conclude, therefore, that feedback does not act as a reward.

Although feedback per se is not a reward, the receipt of feedback indicating that one has saved energy is a potential occasion on which to administer self-reinforcement. Self-reinforcement may range from the spontaneous emotion of happiness that one has achieved or come closer to one's goal of saving energy, through self-congratulatory and self-flattering thoughts and/or statements, to more concrete self-reinforcements—for example, eating a desired dessert. Although there is support for the notion that individuals can regulate their behavior through self-reinforcements (cf. Bandura, 1977b), it remains to be demonstrated that energy conservation behavior is maintained through self-reinforcements.

In our successful feedback experiments (Becker, 1978; Seligman & Darley, 1977), we never offered the subjects an incentive to conserve energy. Why then did they, first, agree to try to conserve and, second, actually conserve? The obvious answer is that the energy saved is the reward. After all, energy savings translate directly into financial savings. However, the money saved must be balanced against the inconvenience and loss of comfort involved in saving energy. Presumably, individuals differ greatly in the way they take these factors into account. But as the literature on obesity (e.g., Bellack, 1975) and smoking control (e.g., Epstein & McCoy, 1975) has demonstrated, the discomfort produced by abstinence is often more powerful than the benefits alleged to accrue from it. Whereas a great number of people have lost weight and quit smoking, another large number of people have found it exceedingly difficult to do so. Whether energy consumption can profitably be analyzed as an addictive behavior like overeating or smoking is a matter for future research. In any case, attempts to manipulate the reward contingencies for a household's energy consumption are going to be difficult. It is unlikely that the price of energy will ever rise high enough to be the sole determiner of energy consumption behavior. Although we have emphasized the cost of energy as a motivation to conserve, we do not mean to slight the other reasons people give for conserving energy—namely, doing a patriotic duty or living according to one's conservation values.

Motivational Approach. Another explanation for the conservation effect produced by feedback posits that feedback has a motivational function, that it facilitates performance by motivating a person to try harder or persist longer at a task. But how does this motivational function operate? Locke, Cartledge, and Koeppel (1968) believe that feedback has a motivational effect because it leads people implicitly or explicitly to set goals for themselves that they then try to achieve. However, little attention has been paid to why people want to improve their performance in the first place. Researchers have assumed some level of initial motivation on the subjects' part to perform the task well, regardless of whether this motivation is an artifact of the experimental setting (Rosenthal & Rosnow, 1969), derives from a basic desire to be competent (White, 1959), or is based on expectations of personal efficacy (Bandura, 1977a). It is highly unlikely that feedback in itself is motivating in the sense of supplying a person with the drive to conserve energy. As Locke and Bryan (1969) point out, feedback is information and thus by itself cannot initiate action. Indeed it would be very surprising to find that simply giving energy consumption feedback to someone with no real desire to conserve could turn that person into an ardent conservationist. Feedback, therefore, is not a primary motivator of energy conservation behavior. However, we have no argument with the suggestion that feedback can help an individual set more realistic goals or help him or her fine-tune existing goals. We prefer, though, to think about this in terms of feedback as goal-relevant information and not as a motivational component of feedback.

Feedback as Goal-Relevant Information. If, as we believe, feedback does not teach homeowners how to conserve energy, does not reward them for saving energy, and does not motivate them to save energy, why does it work? Given that a person is motivated to conserve, and this is an important precondition, feedback can lead to increased conservation efforts by showing that actual conservation is below the level the person wants to achieve. It is in this way that feedback can trigger increased effort; by informing a person as to the actual level of his or her performance, feedback signals when more effort is needed in order to reach a desired standard of performance. With greater effort directed toward improving performance, performance often improves. Indeed, the importance of setting a performance goal and the receipt of feedback relevant to that goal have been recognized as basic elements in any self-control process (Kanfer & Karoly, 1972).

Becker (1978) conducted a study to test the joint effects of goal setting and feedback on energy conservation. One hundred families living in identical three-bedroom townhouses were randomly assigned to five groups. Forty households were asked to set a goal of reducing their electricity consumption by 2% over the next month, an easy goal; whereas 40 others were asked to set a goal of reducing their consumption 20%, a difficult goal. Half the households in both the 2% and 20% goal conditions were given feedback three times a week showing where they stood in relation to their goal. The fifth group of 20 households served as a control; these households were not asked to set a conservation goal and did not receive feedback. On the basis of the preceding analysis of how feedback functions to facilitate performance, the only group expected to reduce its consumption substantially was the one that had been asked to set a difficult conservation goal and that received feedback. The other treatment groups were expected to save only small amounts of electricity, if any. The results (see Table 4.3) were as predicted: The difficult-goal-plus-feedback group used 13% less electricity than the control group. As well as being of practical significance, this difference was statistically significant. As for the other treatment groups, the difficult-goal-no-feedback group used only 1.3% less than the control; the easy-goal-plus-feedback group used 4.6% less; and the easy-goal-no-feedback group used 1.2% more than the control. These differences were not statistically signifi-

TABLE 4.3
Average Daily Electricity Consumption (in kwh)
During the Feedback Period

Descriptive Statistics	*Difficult Goal*		*Easy Goal*		
	Feedback	*No Feedback*	*Feedback*	*No Feedback*	*Control*
Adjusted score[a]					
M	34.7	38.0	39.4	40.3	39.9
SD	8.3	3.6	4.1	5.1	4.3

[a]The adjusted score represents average daily consumption during the treatment period adjusted to take into account differences among the groups in pretreatment consumption (from June 1 to August 8, 1976). The adjustment was obtained by the analysis of covariance method.

cant. The results of this study support the view that feedback operates to reduce energy consumption by letting residents know whether they need to increase their efforts to reach their conservation goals. It also demonstrates that residents must have a fairly challenging conservation goal if feedback is to be an effective conservation aid.

It should be noted that given the within-group variation, the correlation between the covariate (namely, pretreatment consumption) and the dependent variable (namely, consumption during the treatment period), the size of the sample, and control-group consumption, the study could not detect a 2% reduction in electricity use (compared to the control) as statistically significant. However, with respect to the hypothesis under investigation, it was not important whether a small effect was statistically significant. Indeed, statistical significance could have been guaranteed for even a small effect simply by using a sufficiently large number of households in the study. Because the study by Seligman and Darley (1977) found about a 10% reduction in consumption without using a specific goal-setting procedure, support for the present hypothesis required that an effect about as large (or, better, larger) be found in the difficult-goal-plus-feedback group. The present study was sufficiently powerful to detect an effect of this size. In fact, a reduction in consumption of about 8% would have been detected as statistically significant.

Additional insights into how feedback can lead to energy conservation may be obtained by further examining the difference between the feedback and no-feedback groups within each level of goal difficulty. Of the two groups given the difficult goal (conserving 20%), the one receiving feedback used 13.0% less electricity than the control group, whereas the one not receiving feedback used only 1.3% less. When a goal is difficult, a good deal of effort must be made to achieve it. Even without feedback, people know this to be true and, consequently, may exert themselves to an unusual degree. However, even though they may be trying harder than usual to perform well, without feedback, there is no way for them to know whether their efforts are sufficient to lead to the desired high level of performance. Because the effort required to perform at levels far above those that are familiar is usually stressful (or in some way aversive), a pressure constantly exists to drop back to easier levels of effort (and performance). This may lead to an overestimation of the amount of effort actually expended. For example, people new to jogging are often surprised to learn that the very fatiguing two miles they estimated they just ran were really only a mile and a half. Without feedback, it is easier for people with difficult goals to maintain counterproductive misperceptions about the amount or adequacy of their efforts. This may lead to their putting out less effort than that really needed and, consequently, to their performing less well than they are able or want to perform.

Of the two groups given the easy goal (conserving 2%), the one receiving feedback used 4.6% less electricity than the control group, whereas the one not receiving feedback used 1.2% more than the control. In a follow-up questionnaire given after the study, 4 of the 20 households in the 2%-goal-plus-feedback group admitted to having set a more difficult goal after discovering how relatively easy it was for them to achieve a 2% reduction in consumption. None of the households in the 2%-goal-no-feedback group reported setting a higher goal for itself (see Becker, 1977). This finding illustrates another way in which feedback may lead to increased conservation. Feedback may demonstrate to those with only a tentative commitment to energy conservation that saving energy is not as difficult or painful as they may have thought and may thereby encourage them to conserve more.

Feedback may also help residents to maintain their conservation commitment when the going gets rough. Saving energy is similar in some ways to trying to lose weight, quit smoking, and the like. It is easy to make the commitment when we are comfortable or satisfied; it is difficult to keep it when we are faced with repeated periods of discomfort or dissatisfaction. Except for the most determined, many of us need some help in getting through those periods of discomfort without abandoning the goal. Feedback helps keep salient the idea of energy conservation, frequently shows residents how well they are doing with respect to their goals (thereby creating frequent occasions when they can feel satisfied with their efforts) or how poorly they are doing (thereby providing early warning signals that more effort is required), and makes the chore of energy conservation more like a game.

One important issue that future research will need to address is the persistence of energy conservation behavior facilitated by feedback procedures. During the time that feedback is given, how long will energy conservation behavior be maintained? Whereas our treatment periods have been around 3 weeks, other researchers have extended the feedback period to 6 weeks, with successful results (Winett, Neale, Williams, Yokley, & Kauder, in press). Unlike addictive behaviors, like cigarette smoking, which one tries to eliminate for all time, wasteful energy consumption is greatest during certain periods, namely, the cooling and heating seasons. There are many months of the year when space-cooling and -heating requirements are nonexistent or minimal, although of course this varies from one region of the country to another. The point is that for the Northeast, the cooling season is about 10 to 12 weeks, so feedback need only be effective for that amount of time. Feedback has already been shown to be effective for 6 weeks; thus it is plausible that it would be effective for 10 to 12 weeks.

A concern with how long feedback effects will persist after feedback has ceased is of limited practical importance, because there is no reason to stop giving feedback. During peak energy-use seasons, it has been estimated (Winett & Neale, 1979) that it costs less to provide feedback to homeowners than the return in dollars of energy saved. As energy costs are rising more quickly than labor costs, the economics seem favorable to feedback. Additionally, it is likely that devices can be built to present energy consumption feedback in a display located in the home. As a result of the recent and rapid development of microprocessors, it is by no means fanciful to imagine an energy control system that gives homeowners detailed information about the energy performance of their houses and recommends specific energy-conserving actions. And, finally, it has been shown that people are even willing to go through the trouble of reading their own electric meters daily to monitor their energy consumption. An experiment by Winett, Neale, and Grier (1979) demonstrated energy savings of about 7% using self-monitoring techniques.

Feedback and the Hawthorne Effect

The question has sometimes been raised as to whether the effect found in energy conservation studies of the type already described is really a feedback effect at all or, rather, a Hawthorne effect. The argument is that the households receiving feedback are responding not so much to the feedback per se but to the implicit demands of the experiments to conserve and to the increased attention focused on them. There are a number of points to be raised in response to this argument. Often, in studies finding feedback effects, conscious efforts have been made by the experimenters to make demand characteristics and attention effects the same for both the treatment and control groups. In the Seligman and Darley (1977) study, for example, there were no implicit conservation demands. All demands, or rather requests, to conserve were explicit and were identical for the treatment and control groups. In a study by Becker (1978), the amount of attention paid to the groups was the same. All households had charts that were marked with equal frequency, although what the charts indicated was different depending on the experimental condition.

The fact that feedback effects were not always found also suggests that the results of the successful studies were not simply a matter of attention effects or demand characteristics. Becker and Seligman (1978) used procedures similar to those used by Seligman and Darley (1977) and by Becker (1978) in terms of the attention that was focused on the households and experimenter appeals for conservation. Nonetheless, no feedback effect was found in that study, unlike the other studies. Problems in the way predicted consumption was calculated led to feedback that the subjects found difficult to believe. Consequently, many disregarded it. If the feedback effects in energy conservation studies were nothing more than Hawthorne effects, such an effect could still be expected in this study, because the demand characteristics and the attention paid to the households were not affected by the values presented to the subjects as feedback. However, as stated, no such effect was found.

The subjects in our studies have been "real" people living in the "real" world, not college sophomores spending an hour in a laboratory. To conserve energy in the home by behavioral means—for example, turning off the air-conditioning—involves some degree of discomfort and inconvenience. Knowing how difficult it is to get residents to conserve, it is difficult for us to believe that our subjects put themselves out in this way for weeks at a time simply because they thought we expected them to or because they thought it would make us happy.

Factors Producing Effective Feedback

Under what conditions will feedback be successful in facilitating residential energy conservation? Most importantly, feedback must be credible (Becker & Seligman, 1978). By credible, we mean that homeowners should see a rough relationship between their feedback scores and their conservation behaviors. On one hand, homeowners must be explicitly informed about what energy savings are possible from what conservation actions. We must avoid situations where, for example, the homeowner expects 20% savings from turning off lights in unoccupied rooms. On the other hand, we want to avoid abrupt changes in feedback from one presentation to another. Here is a situation where accuracy in feedback might be harmful: A change from 30% saved on one day to 30% wasted on the next may be considered unbelievable even if it represented a true change in the homeowner's behavior. We need to find ways around this problem, perhaps by averaging feedback over several days or by eliminating precise quantitative feedback altogether and presenting, instead, a series of colored lights indicating a few different levels of wasting and saving.

Second, homeowners must be initially committed to saving energy. It is unlikely that feedback can motivate those who are unconcerned. This should not be taken as a limiting condition to feedback any more than it is a limiting condition to any program. Procedures like goal setting (Becker, 1978) or public declarations (Pallak & Cummings, 1976) can be used to enhance commitment.

Third, the frequency with which feedback must be given is probably dependent on both the homeowners' degree of initial commitment to saving energy and their reported difficulty in maintaining conservation behaviors. Probably, the more frequently feedback is given, the better, although Hayes and Cone (in press) have done a study in which monthly feedback was found to be effective. Feedback is probably like chicken soup: It may not always help, but it shouldn't hurt.

Fourth, feedback should be given along a dimension that is most meaningful to the homeowners. Most studies of feedback have used feedback that informed homeowners how well they were doing compared to themselves at an earlier time. Would feedback that compared homeowners to other, similar homeowners be as effective? It is an empirical question whether self- or social feedback is differentially effective for different groups of homeowners.

Fifth, it does not matter how predicted use (of which feedback is a function) is determined, provided the method maintains the credibility and rough accuracy of the feedback. To date, three different methods of calculating predicted consumption have been tried: (1) Predicting a household's consumption from outside temperature on the basis of the regression of consumption on temperature in an earlier period (Seligman & Darley, 1977). (2) Predicting a household's consumption on the basis of a control-group correction. This involves adjusting each household's consumption in the feedback group by the median change in consumption in the control group from the prefeedback period to the present (Becker, 1978). (3) Predicting a household's consumption from outside temprature based on the computation of consumption per degree-day in an earlier period (Becker & Seligman, 1978).

Sixth, the optimal display for presenting feedback is not known. Researchers have typically recorded feedback in a letter given to homeowners or posted the feedback in a chart taped to the homeowner's kitchen window. Hopefully, suitable and economical devices will be built to display feedback. It is also an empirical question whether feedback should be displayed over time so that homeowners will have a record of their performance.

SOCIAL SCIENCE AND ENERGY CONSERVATION

The present research has demonstrated that modest efforts on the part of a household can, under certain circumstances, result in substantial energy savings in the neighborhood of 10% to 15%. This is a significant result in two senses: First, it is statistically reliable; and second, it is important if projected into the traditional metric of energy analysis, which is barrels of imported oil saved. Of course, any full-scale set of solutions to the national energy problem must integrate a web of complex considerations such as utility company financing and regulating, the economics and politics of alternative sources of energy, and so on. Still, the present research argues for sustained attention to the individual level of analysis. Our message is an optimistic one: The individual can play a significant role in solving the energy crisis. And, importantly, as individuals, we can begin to conserve now, without waiting years for technology to produce safe, alternative sources of energy. If we take seriously the idea that individuals, along with more energy-efficient machines, can contribute to energy conservation, then where do we begin? There are several approaches: The most direct implication of our feedback research is that residents who are interested in reducing their energy use should receive the feedback that will facilitate their conservation efforts. We have already outlined several factors that need to be considered in providing effective feedback. But providing people with feedback is part of the more general problem of how to teach people to help themselves to save energy. There are weight-loss clinics, programs for smokers, and self-help organizations like Alcoholics Anonymous; is it reasonable to think in terms of clinics for energy wasters? Effective techniques for saving energy can be disseminated in a number of ways—for example, through mobile energy conservation vans that travel from community to community and that are staffed by experts. These experts can provide several services, from teaching homeowners how to examine their houses for energy deficiences (and providing them with proper instrumentation to do so), to making recommendations about how to improve the energy efficiency of their homes, to teaching homeowners how to monitor their consumption, to helping them to set up reasonable but challenging conservation goals and to calculate their own feedback.

Second, we must learn how to persuade individuals to buy the more energy-efficient machines. Research has only just begun to examine how advertising can take into account the energy savings of a product over a number of years versus the cost of the product at the time it is bought (Hutton & Wilkie, 1979). We should not presume, of course, that energy conservation will be accepted once it makes good economic sense. People often do not act in their own best interest, economic or otherwise. Witness the millions of people who smoke cigarettes, risking lung cancer and emphysema. Part of the problem of getting people to adopt energy-conserving devices is how and by whom the device is introduced to them. Some of these issues can be analyzed from the perspective of the theory of diffusion of innovation (Rogers & Havens, 1962). Darley (1978) has begun the task of showing the applicability of the theory to the area of energy conservation.

Finally, we should note that energy conservation need not imply images of people grudgingly reducing their energy usage (and quality of life) because economic circumstances force them to do so. Individuals can find satisfaction in many of the ways they practice conservation. One way of conceiving of the subjects in our feedback experiments is to say that they became interested in and challenged by the task of lowering their energy consumption, and felt satisfaction not previously present when they did so. Other people have reported the same feelings about recycling practices and other conservation techniques. Addressing the applied question of energy conservation, therefore, returns psychologists to a question that was originally thought to be within the domain of psychology but that has largely been ignored in classic theoretical psychology. What are the values by which a person should live? What actions that a person can take will give that life meaning? It would be ironic if research in an area conventionally thought to be peripheral to the central concerns of psychology would cause us once again to face those central questions.

REFERENCES

Ammons, R. B. Effects of knowledge of performance: A survey and tentative theoretical formulation. *Journal of General Psychology*, 1956, *54*, 279–299.

Bandura, A. Self-efficacy: Toward a unifying theory of behavioral change. *Psychological Review*, 1977, *84*, 191–215. (a)

Bandura, A. *Social learning theory*. Englewood Cliffs, N.J.: Prentice-Hall, 1977. (b)

Becker, L. J. *Reinterpreting the motivational effects of feedback and goal setting on performance: A field study of residential energy conservation.* Unpublished doctoral dissertation, University of California, Davis, 1977.

Becker, L. J. Joint effect of feedback and goal setting on performance: A field study of residential energy conservation. *Journal of Applied Psychology,* 1978, *63,* 428–433.

Becker, L. J., & Seligman, C. Reducing air conditioning waste by signalling it is cool outside. *Personality and Social Psychology Bulletin,* 1978, *4,* 412–415.

Bellack, A. S. Behavior therapy for weight reduction. *Addictive Behaviors,* 1975, *1,* 73–82.

Bilodeau, E. A., & Bilodeau, I. McD. Motor-skills learning. *Annual Review of Psychology,* 1961, *12,* 243–280.

Darley, J. M. Energy conservation techniques as innovations, and their diffusion. *Energy and Buildings,* 1978, *1,* 339–343.

Dorf, R. C. *Energy, resources and policy.* Reading, Mass.: Addison-Wesley, 1978.

Epstein, L. H., & McCoy, J. F. Issues in smoking control. *Addictive Behaviors,* 1975, *1,* 65–72.

Hayes, S. C., & Cone, J. D. Reducing residential electricity use: Payments, information and feedback. *Journal of Applied Behavioral Analysis,* 1977, *10,* 425–435.

Hayes, S. C., & Cone, J. D. Reduction of residential electricity consumption through simple monthly feedback. *Journal of Applied Behavioral Analysis,* in press.

Hutton, R. B., & Wilkie, W. L. *"Life cycle cost": A new form of consumer information.* Unpublished manuscript, School of Business Administration, University of Denver, 1979.

Kanfer, F. H., & Karoly, P. Self-control: A behavioristic excursion into the lion's den. *Behavior Therapy,* 1972, *3,* 398–416.

Kohlenberg, R. J., Phillips, T., & Proctor, W. A behavioral analysis of peaking in residential electricity energy consumption. *Journal of Applied Behavior Analysis,* 1976, *9,* 13–18.

Locke, E. A., & Bryan, J. F. Knowledge of score and goal level as determinants of work rate. *Journal of Applied Psychology,* 1969, *53,* 59–65.

Locke, E. A., Cartledge, N., & Koeppel, J. Motivational effects of knowledge of results: A goal-setting phenomenon? *Psychological Bulletin,* 1968, *70,* 474–485.

McCormick, E. J. *Human engineering* (2nd ed.). New York: McGraw-Hill, 1976.

Pallak, M. S., & Cummings, W. Commitment and voluntary energy conservation. *Personality and Social Psychology Bulletin,* 1976, *2,* 27–30.

Palmer, M. H., Lloyd, M. E., & Lloyd, K. E. An experimental analysis of electricity conservation procedures. *Journal of Applied Behavior Analysis,* 1978, *10,* 665–672.

Rogers, E. M., & Havens, A. E. Predicting innovativeness. *Sociological Inquiry,* 1962, *32,* 34–42.

Rosenthal, R., & Rosnow, R. L. (Eds.). *Artifact in behavioral research.* New York: Academic Press, 1969.

Ross, M. H., & Williams, R. H. Energy efficiency: Our most underated energy resource. *The Bulletin of the Atomic Scientists,* November 1976, 30–38.

Seligman, C., & Darley, J. M. Feedback as a means of decreasing residential energy consumption. *Journal of Applied Psychology,* 1977, *62,* 363–368.

Sinden, F. W. A two-thirds reduction in the space heat requirement of a Twin Rivers townhouse. *Energy and Buildings,* 1978, *1,* 243–260.

Socolow, R. H. The Twin Rivers program on energy conservation in housing: Highlights and conclusions. *Energy and Buildings,* 1978, *1,* 207–242.

Sonderegger, R. C. Movers and stayers: The resident's contribution to variation across houses in energy consumption for space heating. *Energy and Buildings,* 1978, *1,* 313–324.

White, R. W. Motivation reconsidered: The concept of competence. *Psychological Review,* 1959, *66,* 297–333.

Winett, R. A., & Neale, M. S. Psychological framework for energy conservation in buildings: Strategies, outcomes, directions. *Energy and Buildings,* 1979, *2,* 101–116.

Winett, R. A., Neale, M. S., & Grier, H. C. Effects of self-monitoring and feedback on residential electricity consumption. *Journal of Applied Behavior Analysis,* 1979, *12,* 173–184.

Winett, R. A., Neale, M. S., Williams, K., Yokley, J., & Kauder, H. The effects of individual and group feedback on residential electricity consumption: Three replications. *Journal of Environmental Systems,* in press.

Woteki, T. *The Princeton omnibus experiment: Some effects of retrofits on space heating requirements* (Report No. 43). Princeton, N.J.: Princeton University, Center for Energy and Environmental Studies, 1977.

5 Waste Reduction and Resource Recovery: Strategies for Energy Conservation

E. Scott Geller
Virginia Polytechnic Institute and State University

> *"Trash," once put in a can on the curb and forgotten, can no longer be ignored. Solid waste management is now one of the major problems faced by cities and counties throughout the country. As energy costs rise, resources diminish, and land disposal areas become scarce, we must develop economic and environmentally sound approaches for dealing with our municipal solid waste stream.*
> —Community Environmental Council, Inc.

Waste reduction and resource recovery are distinct, waste management approaches for conserving environmental resources and alleviating energy shortages. Waste reduction refers to decreasing the generation of waste by reducing the consumption of environmental resources, whereas resource recovery refers to the extraction of resources from discarded materials for use in manufacturing consumer commodities (cf. Humbler, 1975). More detailed definitions of resource recovery programs (e.g., League of Women Voters of the U.S., 1972) distinguish between using wastes to produce the original product (i.e., recycling) and changing wastes into other materials (i.e., reclamation or conversion). Thus, using waste paper to produce paper, crushed glass to make glass, or scrap iron to produce steel is *recycling* in this narrower sense; whereas burning garbage to generate energy (e.g., refuse-derived fuel) or using crushed glass for road building would be termed *reclamation* or *conversion*. A third term that is sometimes associated with recycling (e.g., League of Women Voters of the U.S., 1972) but more often referred to as waste reduction (e.g., Wahl & Allison, 1975) is *reuse*. Reuse occurs when a product is used over again, usually by another person and sometimes in refurbished form, such as a returnable bottle, a retreaded tire, a dress from Goodwill Industries, or a car radiator from the junkyard. The present discussion has adopted the currently more popular notion that reuse is a form of waste reduction rather than recycling, for it decreases manufacturing energies and thereby attacks the waste management problem at the source (cf. Office of Solid Waste Management Programs, 1975).

Some waste management ''experts'' claim that waste reduction is the only direct method for saving resources and conserving energy; they consider resource recovery to be a costly, post hoc alternative (e.g., Quarles, 1975; Wahl & Allison, 1975; Wendt, 1975). However, other ''experts'' (usually with an alliance to industry) view resource recovery as the only comprehensive, long-term approach to problems of solid waste management and resource conservation (e.g., King, 1975; Van Tine, 1975; Walker, 1975). Actually, an optimal approach to save resources should include both waste reduction and resource recovery strategies (cf. Humber, 1975). For example, packaging materials may not only be reduced (i.e., waste reduction), but much packaging can be produced from recycled paper (James, 1975). Indeed, the president of the Pillsbury Company reported that from 1965 to 1975, his company accomplished a 36% reduction in the amount of metal consumed in packaging (a savings of 11.1 million pounds of metal per year), and each year the company uses about 50 million pounds of recycled paper (Peterson, 1975).

Strategies for resource recovery may be dichotomized as a high-technology or a low-technology system. A *high*-technology recovery system is quite expensive but is most convenient for the consumer. Nonseparated garbage is collected (as usual) and transported to a ''resource recovery'' plant that mechanically separates out paper, glass, aluminum, and metal for reuse in product manufacturing and then burns the remaining refuse with coal and oil to produce steam for electricity generation or oil via a process called pyrolysis (cf. Seldman, 1975, 1976a, 1977). The types and amounts of materials separated from trash before it is processed vary among resource recovery facilities, as does the particular energy conversion process.

The most obvious disadvantage of the high-technology approach to resource recovery is the very high (often underestimated) expense of such a system. For example, the Baltimore Pyrolysis plant was scheduled to open in 1975 at a cost of $16 million, but its operation was delayed because of $10 million in additional costs; and a $70 million project to convert 6000 tons of garbage per day into refuse-derived fuel (RDF) has been planned for St. Louis (LaBreque, 1977). In addition, the maintenance of a high-technology system is quite expensive, as illustrated by the fact that the ''hammermills'' needed to shred the trash in the St. Louis RDF plant cost nearly $5000 for the 6641 tons of garbage processed between November 1973 and April 1974 (''Resource Recovery,'' 1977). Indeed, the annual cost of just operating a high-technology system is substantially higher than the standard landfill approach to solid waste management (Seldman, 1976c).

Other disadvantages or problems with the high-technology approach to resource recovery are as follows:

1. Operating the resource recovery plant consumes much energy and pollutes the air, sometimes with heavy metals ("Resource Recovery," 1977; Seldman, 1975).
2. Using RDF with coal and oil as an energy source causes more air pollution than using coal or oil alone (Seldman, 1975, 1976a).
3. Contracts between resource recovery plants and cities specify some minimum quota of trash each year (below which some cities must pay additional fees) and thereby establish a "disincentive" to reduce waste ("Resource Recovery," 1977; Seldman, 1977).
4. A reliable technology is not available for separating aluminum (the most valuable recyclable) from the solid waste stream, and glass separation is only slightly more advanced than aluminum recovery ("Resource Recovery," 1977).
5. The storage of shredded waste can be a fire hazard ("Resource Recovery," 1977).
6. Locating reliable markets for the energy resource produced by resource recovery plants is sometimes difficult, and market prices vary widely ("Resource Recovery," 1977).
7. Because paper and organics (comprising 55% to 75% of municipal solid waste) are more valuable when recycled than burned for BTUs (cf. Seldman, 1977), increasing amounts of these materials will be removed from the solid waste system via *low*-technology recycling; therefore, the garbage input to justify the operation of *high*-technology systems can be seriously threatened (Seldman, 1976c). In 1976, *low*-technology, source separation programs were operating in more than 75 cities, but 2 years earlier, no such programs existed (Seldman, 1976a).

The *low*-technology approach to resource recovery refers to the more economical but less consumer-convenient system of separating garbage into reusable materials at its source (e.g., the institution, industry, or household that produced the waste) and then transporting separated waste to appropriate manufacturers (e.g., paper to a paper mill, glass to a glass factory, and aluminum to a can company). In other words, the product distribution chain is reversed, and the consumer becomes the first rather than the last link (cf. Margulies, 1970). Thus a *low*-technology system usually requires a mechanism for promoting consumer participation, and determining the most cost-effective and acceptable "promotional" strategy for such a system is becoming an increasingly necessary challenge, especially as community leaders realize the infeasibility of *high*-technology solutions to solid waste management. Behavioral science can certainly contribute to the design, evaluation, and refinement of procedures for promoting communitywide recycling; the present chapter critically reviews those behavioral studies that have provided information for *low*-technology resource recovery.

Before reviewing specific behavioral approaches toward waste reduction and resource recovery, it is instructive to consider the varieties of items that are discarded daily and to examine potential ways of saving the environmental resources used to produce and dispose of our trash. Municipal waste can be categorized into the following five categories: durable waste, nondurable waste, packaging waste, food waste, and yard waste (Wahl & Allison, 1975); each of these trash categories is a potential target for particular waste reduction and/or recycling strategies.

Durable Waste

Durable goods include large appliances (such as televisions, refrigerators, lawn mowers, and furniture) that the consumer decides to replace or discard. The discarded appliances often are damaged beyond repair, but they usually contain valuable scrap materials. Too often, the appliances were discarded because it was more convenient (and rewarding) to purchase a newer model. Because most durable wastes are large, they take up much valuable space in dumps, landfills, and open fields. Automobiles are perhaps the most obvious and unsightly of the durable wastes (as demonstrated by the automobile graveyard).

The extreme difficulty for the average consumer of transporting large durables to repair shops, recycling centers, or scrap yards means that such wastes are often left in garages, backyards, or public spaces. With appropriate advertising and by offering substantial financial rewards, it may be possible to influence the relocation of large, durable wastes; although it might be more practical and energy conserving if recycling centers were to include their own "free" pickup service and offer a small remuneration for consumers who report the location of a recyclable durable and perhaps make an appointment to be on hand when the pickup is made. Indeed, Hayes, Johnson and Cone (1975) suggest that a "marked-item technique" would be a financially practical procedure for getting bulky durables that litter the public environment to recycling centers. Thus, certain durables could be specifically marked unobtrusively by officials from redemption centers, and the reporting and/or appropriate relocation of these marked items would result in a large financial reward. It is noteworthy that large-scale social attention (such as public recognition through the news media) may be as powerful a reinforcer as large sums of money for maintaining the more inconvenient recycling behaviors and may have the further desirable consequence of modeling appropriate recycling behaviors for the newspaper reader, radio listener, and TV viewer.

Tires are considered among the most intractable and indestructible municipal wastes (Kiefer, 1974; Wahl & Allison, 1975)—"not suitable for conventional collection, processing, landfill, or incineration operations'' (Westerman, 1975, p. 84). However, the number of discarded tires would be reduced markedly if more were recycled for producing roofing material, shoe soles, or playground equipment (Seldman, 1976b) and if significant numbers of consumers were persuaded to purchase longer-lasting steel-belted radials. Radials are initially more expensive than standard bias-ply tires, but in fact the total-life cost of the radials is less, because they generally last more than twice as long as standard tires (Wahl & Allison, 1975; Westerman, 1975). Thus, a campaign to promote the purchase of radials would need to convince consumers to delay economic gratification. Perhaps an increased emphasis on the safety advantages of steel-belted radials would promote an increase in waste reduction tire purchases. The use of steel-belted radials might also increase if consumers were provided with economical trade-in incentives. In fact, with at least 1/16 inch of tread remaining, the radial tire could be retreaded to outperform the standard bias-ply tire (cf. Kiefer, 1974).

Nondurable Wastes

A majority of the nondurables are paper-based products (e.g., newspapers, magazines, tissues, towels, paper cups and plates, and so forth), which can be used as a raw material for paper manufacturing (i.e., recycling) or burned as a fuel for energy production. Nondurable goods are also prime candidates for waste reduction programs, since so many nondurables on the market could be replaced with reusables. Indeed, as reusable alternatives for nondurable throwaway commodities are still available (although in decreasing amounts), behavior motivation approaches to waste reduction could attempt to influence the purchase decisions of consumers. Here, again, the convenience of the throwaway is difficult to overcome, even when economics may favor the reusable. Thus, to avoid the inconvenience of washing cloth handkerchiefs, diapers and napkins, dishes and glasses, consumers make the more costly, negatively reinforced response of using paper tissues, disposable diapers, and paper napkins, cups, and plates.[1] Who wants to carry durable, reusable plates and cups to a picnic when a more effortless response is available?

It is noteworthy that attempts to promote the sale of paper dishware often refer to safety and sanitation advantages over glass alternatives. For example, a mother in a 1979 TV advertisement of Dixie Cups refers to the cuts that her child had received from broken glass and how such accidents will be avoided by using paper drink containers. Of course, the viewer is not informed of the availability of safe, reusable plastic cups, nor of the energy waste and waste management

[1]The water and energy needed to wash certain reusables may be more costly than producing compatible disposables; this possibility demands rigorous field testing and extensive cost-effective analyses.

problems caused by large-scale substitution of paper cups for plastic drink containers. Perhaps advertisers should be required to indicate the environmental cost associated with their product. Alternatively (or in addition), it may be warranted to charge the manufacturer (and the consumer) for the waste management costs of producing (and purchasing) nondurables that have durable substitutes. Then in certain cases, it might be possible to offer refunds for the return of durables to recycling centers.

Packaging Wastes

The major portion of municipal waste includes the cans, bottles, cartons, boxes, and wrappings in which consumer goods are transported and sold. In 1971, packaging and containerization consumed approximately 75% of all glass production, 50% of all paper, 29% of all plastics, 14% of all aluminum, and 8% of all steel (Wahl & Allison, 1975; Wendt, 1975). Further, according to Wahl and Allison, just the production of these packaging goods (not including fabrication into containers) required 5% of all the industrial energy used in 1971 (i.e., 1770 trillion BTUs). The unnecessary excess in packaging material is demonstrated by the estimate that if the amount of packaging in the United States *per capita* had been controlled at the 1958 level (the year that marked the beginning of the "packaging explosion"), the country "would have saved more than 566 trillion BTUs, the equivalent of 267,000 barrels of oil each day" (Wahl & Allison, 1975, p. 24).

Governmental controls certainly provide the greatest large-scale impact toward decreasing waste due to unnecessary packaging, but such regulatory litigation usually requires the vote (or social pressure) of the public. For example, to date, public support has enabled governmental restriction of non-returnable drink containers in eight states (i.e., Oregon, Vermont, South Dakota, Maine, Michigan, Iowa, Connecticut, and Delaware) and local bottle regulations in more than 15 communities (e.g., Oberline, Ohio, and Bowie and Howard Counties, Maryland); one state (Minnesota) has enacted a law to regulate new package designs so that they maximize energy conservation and minimize detrimental impact on the environment (Wahl & Allison, 1975). Actually, public support for packaging regulations may depend on the development of community resource conservation projects, and vice versa. Cooperating in one community waste reduction or resource recovery program may not only result in positive attitudes regarding that particular program but may in fact increase the probability of behavioral and attitudinal support for additional efforts toward resource conservation and waste reduction. It is perhaps no coincidence that Oregon was the first state to ban the throwaway drink container (i.e., in 1971); and since 1973, several large-scale, cost-effective and energy-efficient resource recovery systems have been developed and tested in Portland, Oregon (Duncan, 1975, 1976a, 1976b; Seldman, 1975), and a citywide plan has been adopted (Buel, 1977).

Food and Yard Wastes

As early as 1971, several communities were known to be composting yard waste on a relatively large scale (Franz, 1971). Composting consists essentially of stockpiling food and yard wastes and intermittently turning the compost pile to allow oxygenation. The resulting compost (available after 3 weeks under appropriate environmental conditions and at least biweekly turning) can be used to upgrade soil characteristics by: (1) improving workability for plowing and planting; (2) increasing water-holding capacity and thereby increasing resistance to erosion; (3) providing trace minerals and some nutrients; and (4) decreasing soil compactibility so that the soil is better aerated and gas exchange around plant roots is improved (Institute for Local Self-Reliance, 1975). Further, by improving general soil structure and returning organic material to the soil, composting in vegetable gardens has been shown to increase the quantity and nutritional quality of yields. For example, a 9-year study compared the nutrients in potatoes, rye, and oats harvested from composted plots with that from noncomposted (but fertilized) plots and demonstrated 9% more potassium, 6% more nitrogen, and 4% more phosphorus per harvested pound of vegetables in the composted garden (Tietjen & Hart, 1969).

Consequently, composting provides benefits beyond reducing municipal waste (and the energy used to dispose of it), eliminating air pollution from leafburning, and decreasing the leaf clogging of street-drainage systems. And composting requires relatively minimal individual effort—perhaps less effort than bagging food and yard wastes for the garbage collector or piling and burning leaves. Hence for personal-use composting, the members of a household need only include an extra trash receptacle in the kitchen for food wastes and dump the contents on a family compost pile along with leaves and grass clippings. Large-scale composting in urban environments (without sufficient yard space for a local compost pile) can be almost as convenient when curb pickup is available, as in Brookhaven, Long Island, and Maplewood, New Jersey (Franz, 1971). An urban composting program has been planned for Portland, Oregon, whereby a municipal composting plant will combine sludge (processed from human excrement) with food and yard waste that is picked up from households by a recycling firm (Duncan, 1976b). An analysis of energy costs and benefits for recovering and processing various garbage categories demonstrated "that composting has the highest 'net energy' yield of all recycling processes" (Duncan, 1976b, p. 32).

However, before community composting will occur to any great extent, education programs are necessary to teach and promote appropriate composting procedures that minimize financial costs, energy, and individual effort and that meet health standards (e.g., by being relatively free of odors, seepage, and rodents). Moreover, it will probably be necessary to implement community prompting and reinforcement contingencies for the initiation and maintenance of individual or group composting. For example, the city of Brookhaven encouraged neighborhood composting by: (1) distributing free booklets describing procedures and advantages of composting; (2) setting up local demonstration centers to model appropriate composting techniques; (3) initiating highway beautification projects that required compost for growing trees and shrubs; (4) soliciting the cooperative behaviors of Boy and Girl Scouts; and (5) running "compost-pile-building" contests between Scout troups (Franz, 1971). One feasible contingency that may indeed by very effective in maintaining composting behaviors in some communities would be for sanitation departments to charge residents for *each* garbage bag or can that is collected rather than just for the garbage pickup.

BEHAVIORAL STUDIES OF WASTE REDUCTION AND RECYCLING

Most of the sociobehavioral research related to waste reduction and resource recovery has focused on correlating recycling-related behaviors to attitudes rather than determining community interventions for facilitating waste reduction or recycling behaviors (cf. Hendee, 1971; Wiegel, Woolston, & Gendelman, 1977); but it is likely that focusing on attitude–behavioral relationships has less immediate applied value than does focusing on environment–behavioral relationships. For example, opinions regarding environmental litter have not been related to actual littering behaviors (e.g., Bickman, 1972; Heberlein, 1974; Robinson & Frisch, 1975); and even when significant relationships are found between individual characteristics and environment-related behaviors, the applied relevance for the development of community recycling programs is often obscure. For example, a review of several recent attitude–behavioral studies[2] suggested that the participant in a community antilitter or recycling program is likely to be well educated (Arbuthnot, in press; Weigel, in press); concerned about environmental quality (Levenson, 1974; Steininger & Voegtlin, 1976; Weigel & Newman, 1976); liberal in social, economic, and religious philosophies (Arbuthnot, in press; Weigel, in press); relatively high in occupational status (Weigel, in press); and relatively pessimistic about the improvability of environmental quality (Levenson, 1973).

Such correlational data do suggest that educating the public regarding environmental concerns may facilitate community involvement in recycling projects and that the cooperation of liberal, well-educated individuals at high socioeconomic levels will be relatively easy to gain. However, considering environment–behavior rather than attitude–behavior correlations is probably more beneficial for determining ways to motivate public cooperation in commu-

[2]All of these studies related attitudinal measurements to environmental cleanup or recycling; many other investigators have attempted to correlate personality and demographic factors with environmental attitudes.

nity reclamation programs. The need to examine such behavioral research is especially critical, because simple education and advertising procedures have been relatively unsuccessful in encouraging participation in paper-recycling programs (Couch, Garber, & Karpus, 1979; Geller, Chaffee, & Ingram, 1975; Hamad, Cooper, & Semb, 1977; Ingram & Geller, 1975; Witmer & Geller, 1976). For example, our field experiments have shown that a successful paper-recycling program requires the application of individual or group reinforcement contingencies, even when the potential participants are large numbers of college students who should be quite aware of environmental/ecological concerns (e.g., Geller et al., 1975; Ingram & Geller, 1975; Witmer & Geller, 1976).

Table 5.1 summarizes the field studies that applied behavioral strategies to promote waste reduction or resource recovery. Each behavioral strategy that was implemented to motivate responses related to waste reduction and recycling was categorized as an antecedent strategy (i.e., a stimulus preceding the opportunity to make the desired response) or a consequence strategy (i.e., a pleasant event following the desired response). As depicted in Table 5.1, most studies compared the impact between two or more behavioral change strategies, although several interventions included a manipulation of both antecedents and consequences of the target behavior. Indeed, the implementation of a program to reward desirable behaviors in field (or community) settings requires a particular antecedent strategy for announcing the contingency, although some investigators did not specify the nature of the antecedent strategy used to promote their consequence strategy. The footnotes in Table 5.1 summarize critical drawbacks of certain studies; the remainder of this chapter focuses on presenting more particulars of the studies outlined in Table 5.1, including their noteworthy advantages and disadvantages.

Waste Reduction Interventions

Specific procedures for reducing waste at its source are numerous, including: the use of reusable rather than throwaway products (e.g., returnable rather than throwaway drink containers, plastic rather than paper cups and plates, cloth napkins and handkerchiefs rather than paper tissues), the restriction of packaging materials used in shipping and selling products, the purchase of used rather than new commodities (such as clothing from Goodwill Industries, a car part from the junkyard, or a retreaded tire), the purchase of longer-lasting and safer steel-belted radial tires, the production of packages that can be used for additional purposes (e.g., a glass container that can serve as a drinking glass, drink storage container, or a canning jar), the repair rather than replacement of inoperative appliances, and the remodeling rather than destruction of old homes.

However, waste *production* decisions usually have more immediate payoffs than waste *reduction* decisions in terms of response convenience or economics, and therefore the promotion of waste reduction often requires the application of

TABLE 5.1
Summary of Contingency Management Strategies for Resource Conservation

Behavioral Community Research	Antecedents			Consequences		
	Target	Strategy	Outcome	Target	Strategy	Outcome
Geller et al. (1971, 1973) (grocery store)	Purchase of returnable bottles	Flyers distributed at entrance to grocery store	20% increase			
Geller et al. (1975) (college dorms)[a]	Recyclable paper	Signs and posters in dormitory	141 lbs/week 2.2% partic.			
			combined with consequences (cwc)	Recyclable paper	$15 for most lbs. per capita	237 lbs/ week, 3.9% partic.
				Recyclable paper	Raffle coupon per visit[b]	253 lbs./week, 7.3% partic.
Ingram & Geller (1975) (college dorms)[a]	Recyclable paper	Flyer per residence	58 lbs./week, 2.9% partic.			
		Personal delivery of flyer	47 lbs./week 4.3% partic.			
			cwc	Recyclable paper	Raffle coupon per lb.	113 lbs./week, 4.9% partic.
Witmer & Geller (1976) (college dorms)[a]	Recyclable paper	Weekly flyer per residence	49 lbs./week 2.6% partic.			
			cwc	Recyclable paper	$15 for most lbs. per capita	544 lbs./week 5.9% partic.
				Recyclable paper	Raffle coupon per lb.	820 lbs/week 12.2% partic.

(Continued)

TABLE 5.1
(Continued)

Behavioral Community Research	Antecedents Target	Antecedents Strategy	Antecedents Outcome	Consequences Target	Consequences Strategy	Consequences Outcome
Couch et al. (1979) (College dorms)[a]	Recyclable newspapers	Flyer per residence	cwc	Recyclable newspapers	Raffle coupon/ 1/2 lb.	764 lbs./week +470%
Reid et al. (1978) (apartment complexes)	Recyclable newspapers	"Recycling box" in laundry room; signs	7.6 avg. lbs./ day			
		"Recycling box" near dumpsters; signs; door-to-door appeals	13.2 avg. lbs./ day; 74% increase			
Luyben & Bailey (in press) (trailer parks)	Recyclable newspapers	Recycle boxes, flyer/residence	175 lbs./week, 52% increase			
		Posters, flyers, recycling vehicle[c]	cwc	Recyclable newspapers	2 lbs. = 10¢-20¢ toy; 4 lbs. = 20¢-30¢ toy; 8 lbs. = 30¢-50¢ toy	414 lbs./week; 108% increase
Hamad et al. (1977) (elementary school)[c]	Recyclable newspapers	In-class, verbal prompt	198 lbs./week			
			cwc	Recyclable newspapers	Between = class contest, public feedback	2600 lbs./week
Hamad et al. (1979) (elementary school)	Recyclable newspapers	Two "newspaper recycling boxes," general memo		Recyclable newspapers	Public feedback, tree emblem per 120 lbs. of paper	-33% from base-line (157 lbs./ day), 2% partic.

(Continued)

TABLE 5.1
(Continued)

Behavioral Community Research	Antecedents			Consequences		
	Target	Strategy	Outcome	Target	Strategy	Outcome
		Goal-setting memo	cwc		Small trees for all pupils	+56% (366 lbs./day), 2% partic.
		Written and verbal instructions			10-minute extra recess for 75% partic.	+26% (296 lbs/day), 30% partic.
Humphrey et al. (in press) (administrative offices)	High-grade paper	Two wastebaskets	% recyclable paper:[d] 92.3%			
		Divided wastebasket	92.7%			
		Public wastebasket & Personal wastebasket	83.6%			
Arbuthnot et al. (1977) (Athens, Ohio)	Partic. in community recycling	1)Door-to-door survey; 2)Appeal flyer & bag to save cans; 3)Letter & request	% compliance:[e] 1, 2, 3 = 87.5% 2 & 3 = 80% 1 & 2 = 32% 1 & 3 = 20% 2 = 8.2% 1 = 5.9% 3 = 0% controls = 0%			

(Continued)

TABLE 5.1
(Continued)

Behavioral Community Research	Antecedents			Consequences		
	Target	*Strategy*	*Outcome*	*Target*	*Strategy*	*Outcome*
Jacobs (1978) (700 = home neighborhood)	Newspaper recycling (% partic. and inches of paper	Flyer per residence every other week (for 8 weeks)	+ 193% in partic., no change in inches[f]			
			cwc	Newspaper recycling	1¢ per lb.	+120% in partic., no change in inches[f]
			cwc		Coupon for weekly raffle for $5 prize	+390% in partic., no change in inches[f]
Jacobs & Bailey (1979a, 1979b) (Tallahassee-Leon County community)	Partic. in neighborhood program for recycling newspapers and aluminum cans	Brochure and follow-up flyer	Partic. was direct function of SES; no effect of flyer for LM & lower income area			
			cwc	Partic. in recycling program	Pick up on same or diff. day as garb.	+70% for same-day schedule
		Reminder flyer at residence vs. reminder phone call	Slight, transient increase for only door prompt			

(Continued)

TABLE 5.1
(Continued)

	Antecedents			Consequences		
Behavioral Community Research	*Target*	*Strategy*	*Outcome*	*Target*	*Strategy*	*Outcome*
Jacobs & Bailey (1979a, 1979b) (continued)		Newspaper vs newspaper and door-to-door brochure	+200 to 400% due to residential brochure			
	Partic. in neighborhood program for recycling newspapers & aluminum cans	Three-compartment container = RECYCLIT	100% increase			
		RECYCLIT & 2 flyers/week for 5 weeks	+10% with RECYCLIT			
		RECYCLIT, incentive letter, two reminder flyers/ week for 5 weeks	cwc	Partic. in recycling program	Mailbox sticker: "WE RECYCLE"	25% increase, 50% partic.

[a] Poor source for recyclable paper.
[b] Undermining by repeated deliveries with one sheet of paper.
[c] Transportation costs could be excessive.
[d] No data on amount of paper collected.
[e] Participation judged by telephone interview, not behavioral observation.
[f] Cost-benefit analysis indicated that no intervention provided for a self-supporting program

contingencies that offer special rewards for waste reduction strategies. For example, the large-scale waste reduction program implemented by the Red Owl food stores in six Midwestern states provided monetary refunds when shoppers returned certain packaging materials for reuse—that is, 2¢ per shopping bag, 3¢ per egg carton, and 4¢ per refillable milk container (see Greene, 1975, for a more complete discussion of this "bring 'em back, repack and save" program).

Manditory bottle-deposit legislation is the one waste reduction strategy that has been receiving the most attention these days. Amidst much controversy, the first statewide "bottle bill" became effective in Oregon on October 1, 1972. The bill required every beer and soft-drink container sold or offered for sale in Oregon to have a refund value of not less than 5¢, and it forbade the sale of pull-tab drink containers (Savage & Richmond, 1974). The initial Vermont bottle bill (effective in September, 1973) also provided a minimum 5¢ deposit on all beer and soft-drink containers, and a subsequent revision in Vermont's beverage-container law banned nonrefillable glass containers as of January 1, 1977 (Loube, 1975). Moreover, the fall of 1976 marked the initiation of a regulation by the Environmental Protection Agency that all federal facilities (including national parks and forests) handle only beverage containers with a 5¢ deposit (Fowler, 1976); in December 1978, bottle bills similar to those in Oregon and Vermont were implemented in Michigan and Maine, except that in Michigan the mandatory deposit is 10¢ rather than 5¢ (Dernbach, 1977; "Voters Ban Throwaways," 1976).

Some people report "a rip-roaring success" for the manditory deposit legislation (e.g., Ruvin, 1975; Savage & Richmond, 1974; Skinner, 1975; Waggoner, 1976); whereas others insist that the bottle bills have done more harm than good (e.g., King, 1975; Van Tine, 1975). Included among the reported benefits of the Oregon and Vermont bottle bills are the following:

1. A 65% reduction in beverage-container litter for both states (Loube, 1975; Skinner, 1975) and a 35% decrease in all litter categories in Oregon (Savage & Richmond, 1974).
2. A bottle-return rate above 90% and an average of 8 to 12 refills per bottle in Vermont (Loube, 1975) and as high as 18 refills per bottle in Oregon (Savage & Richmond, 1974).
3. An 81% reduction of container purchases by bottlers and brewers in Oregon, thereby reducing solid waste by 350 million containers (Waggoner, 1976).
4. Increased public awareness of waste reduction and recycling issues and consumer acceptance for other resource recovery projects in Oregon (Bree, 1977).
5. A decrease in litter-related warnings and citations in Oregon (Bree, 1977; Waggoner, 1976).
6. A more than 100% increase in human energy and a 50% decrease in environmental energy in the production of drink containers (Savage & Richmond, 1974).

Most of the reported negative ramifications of the bottle bills seem to reflect social and economical adjustments and therefore may be only temporary. For example, in Vermont, price increases during the 1st year of mandatory deposits ranged from 20¢ to 40¢ per case of soft drinks and from 40¢ to 60¢ per case of beer (not counting the 5¢-per-bottle deposit); sales of soft drinks and beer declined significantly (Loube, 1975). Similar effects occurred in Oregon during 1972, but in 1973 (2 years after the initiation of mandatory deposits), sales of soft drinks and beer regained the increasing trend shown throughout the 10 years preceding the Oregon bottle bill (Savage & Richmond, 1974). Although some skilled jobs were lost, there were other, less skilled jobs added (e.g., warehouse sorters, truck loaders, and drivers); in fact, a net increase in employment was reported for Oregon (Ruvin, 1975; Savage & Richmond, 1974). In addition, beverage-related transportation costs increased substantially in Vermont, but they are apparently declining with the use of fewer container types and sizes (Loube, 1975).

Prevailing negative outcomes of the bottle legislation are hassles associated with handling and storing returnables (Loube, 1975; Serrin, 1979), which could also become a health hazard. This latter disadvantage would be eliminated entirely at retail outlets with the adoption of a community resource recovery network (like the ORE Plan outlined by Duncan, 1976b), whereby residents leave empty returnable bottles with their garbage for pickup by recyclers, who then deliver the returnables directly to bottlers for reuse. The consumers' bottle-deposit refunds could be subtracted from the cost of the garbage pickup.

The cost-effectiveness of mandatory deposit legislation requires a careful, long-term examination of many variables that change with a large-scale shift from throwaways to returnables. However, the environmental/ecological advantages of the bottle bills have been overestimated by environmentalists and underestimated by industrialists. Environmentalists claim a net energy savings and substantial litter reductions from the reuse of drink containers, whereas industrialists argue that the litter counts are biased and nonrepresentative and estimate that the extra energy costs to transport and sterilize reusable containers actually make throwaways more energy-efficient than returnables. Obviously, there is an ent need for objective, unbiased data regarding the energy, litter, economic, and social consequences of banning the throwaway drink container. As of this writing, eight states had enacted laws to ban throwaways (Serrin, 1979), so opportunities are certainly available for evaluating objectively the short- and long-term costs and benefits of such legislation.

Field studies to influence grocery store customers to purchase their soft drinks in returnable rather than throwaway containers represented the first systematic attempt to apply principles of behavioral analysis for waste reduction (Geller, Farris, & Post, 1973; Geller, Wylie, & Farris, 1971). For both of these investigations, the dependent variable was the proportion of those customers who purchased soft drinks available in both returnable and nonreturnable bottles who selected returnable bottles for more than 50% of their purchases. Frequency data for calculating daily proportions were recorded by behavioral observers posted at the stores' checkout counters. As summarized in Table 5.1, the basic intervention for both projects was a simple antecedent strategy, whereby all customers entering the grocery store during the observation period were handed a one-page flyer stating the desired response and listing reasons for the request.

Geller et al. (1971) used an ABA design to study the effectiveness of waste reduction prompting in three supermarkets. That is, at three grocery stores, baseline measurements were obtained during daily 1-hour sessions for 6 consecutive weekdays (including Saturday), and then the prompting intervention was implemented during the same 1-hour period each weekday for 2 weeks. Follow-up data (with the intervention removed) were obtained immediately following treatment (for 6 consecutive weekdays) and then for 1 month after treatment (for 2 consecutive weeks). A problem with this experiment (and with the follow-up study by Geller et al., 1973) was that the same individual's behavior could not be observed under different conditions; therefore, the actual frequency of individuals influenced by the environmental interventions could not be measured. In a sense, the milieu (in this case, the grocery store) was the "organism" subjected to behavior modification technology (cf. Nietzel, Winett, MacDonald, & Davidson, 1977, p. 322).

The results of our first study (Geller et al., 1971) showed prompting to have no consistent influence over the returnable-bottle purchases in two large supermarkets (i.e., Krogers and Mick-or-Mack); but in a small, quick-stop store where customers often purchased only one drink (i.e., Seven-Eleven), the proportion of drink customers who purchased more than 50% of their drinks in returnables was higher on each treatment day than the highest baseline or follow-up day. The mean percentage increase in returnable-bottle customers at the Seven-Eleven store was 32%. Our eventual interpretation of the between-store differences was based on economic influence. That is, each week certain drinks were on sale in the larger supermarket (whereas sales never occurred in Seven-Eleven), and drink customers were more apt to be influenced by money savings than by waste reduction prompting. This interpretation was in fact supported by a finer data analysis that correlated the particular drink flavors selected with price fluctuations, and by unpublished research that was accomplished in preparation for the Geller et al. (1973) experiment. That is, regardless of their containers, drinks that were reduced in cost for a particular week were consistently favored.

The follow-up study (Geller et al., 1973) took place at the quick-stop Seven-Eleven store where the prices of soft drinks were stable from week to week. Over a 4-week period, six 2-hour experimental conditions were varied within weekdays (including Saturdays) according to a Latin-square design. More specifically, the customers' bottle purchases were recorded for a baseline condition and for the following variations of response priming (or prompting): (1) one college student gave incoming customers a handbill that urged the purchase of returnable bottles by appealing to specific antipollution advantages of returnable containers; (2) one student distributed the handbill, and on a large obtrusive "Pollution Chart" adjacent to the entrance/exit door, another student publicly recorded each bottle customer as a returnable-bottle customer (i.e., purchasing more than 50% returnables) or a throwaway-bottle customer; (3) one student distributed the handbill while a five-member, mixed-sex group charted each exiting bottle customer; (4) one male student distributed the handbill, and three male students did the charting; (5) a female student distributed the handbill while three female students charted the bottle customers.

Each of the five antecedent strategies influenced approximately equivalent increases in the proportion of returnable-bottle customers. More specifically, during baseline, the weekly percentage of returnable-bottle customers ranged from about 50% to 75%, with a mean of 60%, whereas under the prompting treatments, the weekly percentages were never below 60% and reached highs of 90%. The overall mean percentage increase in returnable-bottle customers during treatment was 22.2%. The variations of 2-hour experimental conditions within 24 consecutive weekdays provided much more representative samples than those of the prior study and made it likely that the same drink customer would contribute data for more than one condition. Indeed, the results of 76 interviews conducted during the last week of the project showed that 26% of the customers shopped at the Seven-Eleven store at least six times a week and that 63% of the patrons shopped at the store from one to five times per week. Given the likelihood that many persons were observed under different conditions throughout the 24 consecutive weekdays, it is probable that some individual behavioral changes were reflected by the treatment effects. Actually the possibility of between-session generalization could certainly have masked differential treatment effects, but the crudeness of the dependent measure also limited the extent to which differential effects of antecedent strategies could be detected. Perhaps a ratio of the number of returnables to the number of throwaways purchased per customer (rather than the simple "returnable" versus "nonreturnable" categorization) would have been sensitive to our variations in antecedent change strategies.

Waste Recovery Interventions

Although the market for wastepaper has traditionally been unstable and incentives are not yet available for encouraging major paper mills to use secondary paper on a regular basis (Smith, 1975), wastepaper has been the most frequent target for waste recovery programs. This may be the case for a number of reasons, including the following: (1) Paper is the most convenient recyclable to separate from trash; (2) the abundance of recyclable paper in households and institutions can make even relatively small-scale projects economical and even profitable; and (3) the environmental benefits from recycling paper are obvious. Recycling paper conserves energy (secondary paper requires 60% less energy than virgin material to produce paper), reduces air pollution (secondary paper emits 60% less air pollutants than virgin materials), saves landfill space (by diverting significant portions of solid wastes from municipal garbage), and helps to preserve our forests (''Paper Profits,'' 1977).

Nearly all of the behavioral research related to resource recovery has targeted paper recycling. To the author's knowledge, Geller et al. (1975) published the first account of a systematic attempt to analyze the behavioral effects of particular interventions designed to facilitate a paper-recycling program. For about 4 months preceding this study, which took place on the campus of Virginia Polytechnic Institute and State University, students from the campus Committee for Ecological Rebalance (REBAL) had been operating a universitywide paper-recycling program. Each student dormitory had a small storage room designated as a ''Paper-Collection Area'' (PCA), and colorful posters tacked to the dorm bulletin boards gave the location of the relevant PCA and urged cooperation in order to ''Save Our Trees.''

Every 2 weeks, the students from REBAL gathered and bundled the paper in each dorm's PCA and transported and sold the paper to a local mill for $5 per ton. The 6-week study by Geller et al. was designed to evaluate the efficacy of this recycling project and to define contingencies that would increase student participation in the program.

The three pairs of dormitories in our initial study were matched for sex and number of residents, and each dorm was in close proximity to its matched dorm. Three experimental conditions, lasting 2 weeks each, were alternated within each dorm pair according to a Latin-square design. More specifically, the following conditions were varied so that each 2-week condition occurred once for each dorm pair and at each of the three possible sequential positions (i.e., first, second, and third): (1) *Baseline*—Identical to the preexperimental situation except that two data recorders were stationed at the PCA from 5:00 P.M. to 7:00 P.M., Monday through Saturday. Upon each paper delivery, the data recorders measured and tabulated the weight of the paper and thanked the students for their paticipation. (2) *Contest*—$15 was awarded the dormitory of the pair (males vs. females) that delivered the most recyclable paper per capita in 1 week. A sign on the door to the PCA displayed the specifics of this contingency, including the 2-hour period each day when deliveries would be tallied for the contest. (3) *Raffle*—A numbered ticket for one chance in a weekly raffle of prizes donated by local Blacksburg merchants was given for each delivery of at least one 8-½-by-11-inch sheet of paper or cardboard. The rules for this contingency were posted on the door to the PCA, as was a list of the four prizes that were to be raffled off on the evening of the upcoming Friday.

The results of our study demonstrated a consistent advantage of extrinsic reinforcement contingencies, in that the contest and raffle contingencies were almost equivalently effective in markedly increasing program participation over baseline levels. The total pounds of paper delivered per experimental condition were 845 for baseline, 1420 for the contest, and 1515 for the raffle. The percentage of resident participation was consistently greater during the contest and raffle interventions than during baseline but was lower than expected and desired during all conditions. For example, the mean percent of dormitory residents who delivered at least one 8-½-by-11-inch sheet of paper to the collection room over a 1-week period was 2.21% for baseline, 3.87% for the contest, and 7.31% for the raffle. Geller et al. suggested that the low participation rates were due to ineffective antecedent strategies (i.e., response priming) rather than weak reinforcement contingencies. In other words, relatively few residents may have attended to bulletin board announcements, and therefore a majority of the residents may have been unaware of the recycling program.

A follow-up to the Geller et al. study was accomplished 1 year later in different dormitories and attempted to refine the antecedent intervention of the earlier study with the distribution of flyers to each dormitory room (Witmer & Geller, 1976). For a prompt-only condition, a flyer that announced the location of the dorm's PCA and urged cooperation was slid under the door of each dorm residence on each of 3 consecutive Sundays. The same basic flyer design used for the prompt-only condition was used for the contest and raffle interventions except that in place of the stated rationale for paper recycling, there was a precise definition of a group (contest) or individual (raffle) reinforcement contingency. Another noteworthy change in this follow-up experiment was in the definition of the raffle contingency. Specifically, during our first raffle (Geller et al., 1975), we reinforced each delivery of at least one 8½-by-11-inch sheet of paper to the collection room with a raffle coupon. As we should have anticipated, this contingency resulted in some residents undermining the system by stacking their paper outside the door to the collection room and then repeatedly delivering individual sheets of paper to the data recorder, receiving a raffle ticket for each delivery. The raffle contingency in the Witmer and Geller study did not encourage such repeated deliveries to the collection room, since participants received one raffle ticket for each pound of recyclable paper delivered.

Instead of alternating the experimental conditions in a Latin-square design as Geller et al. (1975) did, Witmer and Geller (1976) took baseline recordings for the same 2 weeks in six college dormitories; then for 3 weeks they implemented a prompting procedure in one pair of dorms (males vs. females), a raffle contingency in a second dorm pair (males vs. females), and a contest contingency in the third dorm pair (male civilian vs. cadets). The paper drive and reinforcement contingencies were detailed on the flyer that was placed under each dorm resident's door on each Sunday of 3 consecutive treatment weeks. Then all interventions were removed from the dormitories, and participation data were recorded for 3 additional weeks.

The results of the Witmer and Geller study were quite similar to those of Geller et al., suggesting the need for reinforcement contingencies in a paper-recycling program. Thus, prompting alone (i.e., an antecedent without a consequence) resulted in a total of 147 pounds of recyclable paper in 3 weeks—just barely more than the paper delivered during the 2 weeks preceding the distribution of the flyer (i.e., baseline). On the other hand, the amounts of paper delivered to the collection rooms increased substantially during the 3 weeks of an individual- or group-reinforcement contingency, totaling 2459 pounds for the raffle and 1633 pounds for the contest. The reinforcement contingencies increased the percent of resident participation as well as the pounds of paper collected, but—as in the Geller et al. study—the number of dorm residents who delivered at least one sheet of recyclable paper to the collection center per week was discouragingly low. More specifically, the highest percentage of weekly resident participation occurred for the raffle condition, reaching peaks of 14.4% during the 2nd raffle week for females and during the 3rd raffle week for males. The overall mean participation *per week* were only slightly higher than in the Geller et al. study—averaging 1.89% per week during baseline, 2.60% during prompt-only, 5.85% during the contest, and 12.22% during the raffle. Witmer and Geller speculated that the contest contingency was only markedly effective in the civilian dorm because only these residents had specified a method for spending the contest winnings. Indeed, the civilians planned a dorm party and used their total winnings of $45 to procure beer.

The percentages of dorm residents who participated in the paper-recyling programs of Geller et al. (1975) and of Witmer and Geller (1976) were similar and disappointingly low (i.e., never greater than 10% participation per day). As every dorm resident in the Witmer and Geller program received three flyers announcing the paper-recycling program, it is doubtful that low participation was due to students not becoming informed of the recycling program (as had been speculated by Geller et al., 1975). Apparently, low participation rates resulted from lack of contingency effectiveness rather than lack of contingency awareness. Indeed, this conclusion was supported by the results of a third paper-recycling study accomplished on the Virginia Tech Campus (Ingram & Geller, 1975). The design and procedures of this 8-week experiment were similar to those of Witmer and Geller, and the results essentially replicated both Geller et al. and Witmer and Geller. One additional intervention in this study involved *personal* presentations of the program flyers to dorm residents. That is, for this verbal-prompting intervention, a "prompt distributor" knocked on the resident's door, handed the resident a flyer similar to that used by Witmer and Geller (1976), and said the following: "I would just like to remind you of the paper drive being held Monday through Friday from 5:30 to 7:30 in the basement." When the raffle condition was in effect, the verbal prompt and flyer included a description of the raffle contingency.

The personal touch provided by verbal prompting in the Ingram and Geller study did not facilitate participation in the paper collection program. More specifically, during the verbal-prompting procedures (without the raffle contingency), the average percentage of residents who delivered paper to the collection room each week was only 4.29%, compared to 2.89% per week when the flyers were merely slid under the residents' doors. The weekly participation percentages averaged only 4.92% during the condition that combined a verbal-prompt-and-raffle contingency and declined to 3.71% during 2 weeks of follow-up when all contingencies were removed (i.e., prompting and raffle ticket reinforcement). The mean pounds of paper collected per week in the Ingram and Geller study was 58.0 for the flyer-prompt-alone condition, 46.7 for the flyer-and-verbal-prompt interventions, and 112.5 for the verbal-prompt-and-raffle contingency.

It is noteworthy that for each of our paper-recycling projects, the raffle contingency was slightly more effective than the contest in encouraging program participation, thereby suggesting an advantage to rewarding individuals rather than groups, at least when the group is large. Also notable is the cost-effectiveness of the raffle and contest contingencies. The contest cost only $15 a week (per dorm pair), and the raffle prizes were donated by community merchants for the publicity accompanying their support of a community ecology project. In fact, as a result of the three projects, store managers were approached on 71 occasions for donations of raffle prizes, and 60 of those requests (i.e., 84.5%) resulted in a contribution of at least one raffle prize, amounting to a grand total of 74 prizes, ranging widely in value (from $1 to $30) and in kind (e.g., among the donated prizes were a pitcher of beer, a styled haircut, a bicycle rack, a wall poster, a submarine sandwich, a bottle of cologne, a record album, a steak dinner, and a sleeping bag). Furthermore, six different newspaper stories reported aspects of the paper drives throughout the 2-year period; all contributing merchants were mentioned at least once, and on several occasions a picture of a store manager handing a prize to a raffle winner appeared in the newspaper. These examples demonstrate the potential community cooperation available if appropriate contingencies are implemented. The only real cost is the time and effort to organize, implement, and evaluate such large-scale, contingency managment procedures.

A paper-recycling project similar to our projects was completed recently by Couch, Garber, and Karpus (1979) on another university campus in Virginia (i.e., at James Madison University). After 1 week of recording baseline frequencies of daily paper deliveries and weight of paper per delivery at two dormitories, these investigators distributed a flyer to each dorm resident that detailed a raffle contingency, whereby a raffle ticket was exchanged for each .5 pound of recyclable paper. After 2 weeks of this raffle intervention, the required paper per raffle ticket was gradually increased in one dormitory over the subsequent 4 weeks, from one ticket per .5 pound of paper to one ticket per 3 pounds of paper. For the other dormitory, the initial high reinforcement density remained the same throughout treatment. Then, after six weekly raffles, the reinforcement contingency was removed in both dorms, and assessments of program participation continued for 1 week of follow-up.

The results of the Couch et al. study indicated prominent increases in paper deliveries as a result of the raffle contingency. For example, the total amount of paper delivered to the collection room was 134 pounds during the first baseline week, averaged 763.7 pounds per week through the 6 weeks of the raffle intervention, and decreased to a between-dorm average of 240 pounds for the last week of the project when the raffle contingency had been removed. Throughout the entire 8-week experiment, 106 different residents delivered some recyclable paper to the dorm collection room, representing 27% participation. Note that this value is based on the entire experiment, whereas we reported participation rates per week. Ingram and Geller (1975) did report the total number of dorm rooms that was represented by the paper deliveries, finding 89 of a possible 443 rooms represented or 20.1% participation. Given that two to three students resided in each of the dorm rooms, this percentage would have been somewhat higher if residents rather than residences had been recorded and would probably have come quite close to the overall participation rate calculated by Couch et al. (i.e., 27%).

The number of students who participated in the four university recycling programs already described was disappointingly low under all conditions. These results certainly suggest that university dormitories are inappropriate environments from which to run a paper-recycling program, perhaps because the average dorm resident does not attend to poster or handbill announcements, because meagerly reinforced recycling behaviors are too inconvenient to fit into the busy behavioral repertoire of the average college student, or because the supply of recyclable paper per dorm resident is perceived as too minimal to make paper deliveries worthwhile.

These disadvantages of a university paper-recycling program were less critical when a paper-recycling contest was implemented and evaluated at an apartment complex (Reid, Luyben, Rawers, & Bailey, 1978). During baseline, a "recycling box" was located in the laundry rooms of three apartment complexes, and a handwritten sign on each laundry room wall identified the recycling box. The proximity-and-prompt condition had two components and was initiated at three apartment complexes at different times from the start of the 20-week study (i.e., after 1 week, 3 weeks, and 27 weeks). The proximity component of this intervention involved the distribution of two additional recycling containers (with identifying signs) adjacent to garbage dumpsters located in the parking lot for each complex (Reid, Luyben, Rawers, & Bailey, 1978). During baseline, a "recycling box" was located in the laundry rooms of three apartment complexes, and a this project indicated prominent and consistent increases in pounds of newspaper collected as a result of the proximity-and-prompt procedure. For example, at one apartment complex (72 residences), the mean pounds of paper collected daily increased from 7.91 during baseline to 16.48 pounds during the proximity-and-prompt intervention. These daily means were 9.16 versus 14.14 pounds at another apartment complex (48 residences) and 5.59 versus 8.99 pounds at the third complex (64 residences). An obvious advantage of this study, compared to other paper-recycling programs, was the "inexpensiveness and relatively small amount of person power required to achieve these results" (Reid et al., 1978, p. 9).

A paper-recycling project implemented by Luyben and Bailey (in press) applied a positive reinforcement procedure unlike those reviewed earlier. For a baseline condition, these researchers placed a large plywood box at the entrance to each of four trailer parks in Tallahassee, Florida, conveying the message: "Please recycle newspapers here." After 4 weeks of baseline at one trailer court and 7 weeks of baseline at another trailer court, a prize-and-prompt intervention was initiated. This intervention involved the following procedures: (1) mounting a poster in the park's laundromat and distributing a flyer to each mobile home that announced the beginning of a "Newspaper Recycling Reward Program" for children in the trailer park; (2) distributing a registration card with each flyer that parents were requested to sign in order to give permission for their child to participate in the reward program; (3) driving a "recycling vehicle" to the park at 7:00 P.M. on designated collection days, weighing child-delivered newspapers, and then offering participants their choice of a toy within a particular price range. The price range was dependent on how much paper the child had delivered—a toy selection from a 10¢ to 20¢ category for 2 pounds of paper, and a selection from a 30¢ to 50¢ category for 8 pounds of paper. As a result of the prize-and-prompt intervention, the average pounds of newspapers collected per week increased from 242 during baseline to 616 pounds at one trailer park and from 132 to 212 pounds at the second trailer park. Removal of the prize-and-prompt condition after 8 weeks resulted in returns to the lower baseline levels at each trailer court.

The prize-and-prompt procedure administered by Luyben and Bailey was analogous to the intervention used previously by other researchers for motivating children to gather litter from environments (e.g., Burgess, Clark, & Hendee, 1971; Chapman & Risley, 1974; Clark, Burgess, & Hendee, 1972) and suffered from a weakness that also plagued these antilitter studies. That is, the antilitter contingencies that gave children inexpensive rewards for bags of litter set the occasion for the undesirable behaviors of filling litter bags from sources of already collected litter; similarly, rewarding children for used newspapers could have resulted in the collection of unused newspapers from front yards or porches before they were retrieved by residents. Also, it is conceivable that such a contingency could work against waste reduction by influencing some children to nag their parents and friends to buy "extra" newspapers that they would not normally purchase. An additional undesirable side effect could have occurred as a result of the Luyben and Bailey price-scaling procedure that offered a more expensive toy for more paper. Specifically, this contingency could have promoted the accumulation of recyclable newspapers in children's trailers (perhaps with parental annoyance) until the pile was high enough to permit selection from among the most expensive toys. Hence, a simple positive reinforcement procedure can encourage a variety of undesirable behaviors; such behaviors may not be anticipated prior to program implementation and must therefore receive attention during continual program evaluation.

Hamad, Cooper, and Semb (1977) organized a paper-recycling contest in a public school (Grades 1 to 6), whereby a coupon redeemable for a hamburger was given to each participating student in the one classroom out of three or four that collected the most paper in 3 ½ weeks. Then the contest contingency was dropped for 2 weeks, but the paper drive was still in effect as announced in class and on written memos that were given to the pupils to take home. During the initial contest, 11,254 pounds of paper were delivered to the collection site (i.e., a garage adjacent to the school grounds), but only 292 pounds of recyclable paper were delivered after the contest was dropped. When the contest contingency was reinstated for 2 weeks, 3044 pounds of paper were collected; and during the next 2 weeks (when verbal appeals were given in class but no extrinsic rewards were offered), the total pounds of collected paper were only 500. During the two contests *only*, daily feedback of the paper collected by each classroom was displayed on a thermometer-type scale in a hallway of the school. Thus, the contest included *two* major changes from the baseline or verbal-prompt condition (i.e., a group competition and public feedback); either one or both of these components were necessary in effecting the behavioral differences. The authors reported that 136 of the 271 potential participants (i.e., 50.2%) delivered paper at least once, but they did not give participation rates per experimental condition.

The rate of participation and the quantity of paper collected were markedly greater in the Hamad et al. program than in any of the documented paper-recycling studies that had been established in college dormitories (cf. Couch et al., 1979; Geller et al., 1975; Ingram & Geller, 1975; Witmer & Geller, 1976). At least two factors contributed to this difference, and each should certainly be considered when developing recycling programs. First, it is probable that more elementary pupils than college students became aware of the recycling program implemented by Hamad et al., because in the elementary school, prompting occurred in the classrooms when the children were more or less a captive audience. Further, the parents of the grade-school pupils were prompted to join the recycling effort when the written memos reached their intended destinations. Perhaps the factor contributing most to the superiority of paper recycling in an elementary school versus a college dormitory was that a much greater source of recyclable paper was tapped with the elementary-school program. The supply of paper materials is usually substantially higher in a child's home than in a dormitory room. The university programs would have collected more paper if they had targeted more plentiful sources for recyclable paper, such as faculty and administrative offices, publication services, and computer centers. Such was indeed the case in the program developed by Humphrey, Bord, Hammond, and Mann (in press), which is summarized later in this chapter.

The relatively high rate of participation in the Hamad et al. study compared to those of other recycling programs that attempted to motivate participation with a contest may be partially attributed to a critical distinction between the contest contingency defined by Hamad et al. and those defined by Geller et al. and Witmer and Geller. For both contingencies, the reward was dependent on the performance of the whole group, but the Hamad et al. contest involved an individual component not usually included in group contingencies. That is, only those pupils who delivered at least one sheet of paper to the collection site received the reward *if* their class won the contest. The dormitory contests administered by Geller and his students awarded the whole dormitory $15, and the dormitory to collect the most recyclable paper was the one whose residents specified a method for spending the contest winnings. Every resident in the winning dorm was permitted to enjoy the group reward (i.e., beer), which was purchased with the winnings. More contest participation might indeed have occurred in the dormitories if the contest contingency had been more like that defined by Hamad et al., such that only those residents who made a paper delivery could indulge in the beer drinking.

An important aspect of the Hamad et al. study was a cost–benefit analysis of their program, which collected 7.5 tons of recyclable paper. Cost–benefit analyses should become standard procedure in the evaluation of waste management programs. The authors considered four components in their analysis: (1) revenue from the sale of reclaimed materials = $148.42 earned, at approximately $20 per ton; (2) reduced solid waste = $166.00 saved, considering a national cost of $22 per ton for collecting, transporting, and disposing of solid waste (Environmental Protection Agency, 1974); (3) energy from the reuse of materials = $101.15 saved if the paper was burned as fuel, or $147.00 saved if the recycled fibers replaced virgin wood pulp in manufacturing paper; and (4) natural resource savings = the utility of not depleting natural resources for energy and/or product manufacturing. Weighing the first three benefits against the costs of the hamburgers ($47.24) and the 153 miles of truck transportation ($30.60) results in a net monthly gain of $337.72 for the short-term paper drive. However, the costs of managing the paper collection center (53.5 hours) and transporting the paper to school (which the authors did not specify) must be considered in a complete cost–benefit analysis.

The cost of transporting the paper to the school when this was accomplished by a "special" trip with a motor vehicle was a noteworthy energy cost for the Hamad et al. project. Since 1 of the 2 days that the collection garage was open each week was Saturday, it is possible that a high percentage of the total 240 car or truck deliveries involved a transportation cost that served no other purpose but to deliver paper. Such energy costs for special transportation could be eliminated if the collection center were only open when parents normally drive their children to school or, better yet, if contingencies were contrived to encourage children to carry small amounts of recyclable paper with them when walking to school each morning. Indeed, an advantage of developing recycling programs in public schools is that recyclable materials, readily available in the home, can ride "piggyback" with students and thus contribute no costs due to extra transportation.

One year following the initial Hamad et al. study, these authors conducted a second study at the same elementary school that did not promote special automobile trips for paper deliveries (Hamad, Bettinger, Cooper, & Semb, 1979). Specifically, two large plywood boxes labeled "Newspaper Recycling" were placed at the school entrance and were locked shut except during the collection period from 8:00 to 8:30 A.M. each schoolday. Because school began at 8:30 A.M., this collection procedure set the occasion for paper deliveries without extra transportation costs. Hamad et al. (1979) measured pounds of paper and frequency of participation during seven conditions, sequenced as follows:

1. *Baseline 1* (approximately 9 weeks). Students took home a memo from the school principal that detailed the paper collection times, indicated that proceeds would be donated to the school PTA, and encouraged participation.

2. *Public Feedback* (approximately 6 weeks). A poster displayed publicly the cumulative pounds of paper contributed by each of the 11 classrooms. A small emblem of a tree was stamped next to each thermometer-type scale whenever 120 pounds of paper were collected.[3]

3. *Baseline 2* (approximately 3 weeks). The feedback sign was removed.

4. *Goal Setting* (3 weeks). Students took home a second memo from the principal that introduced a schoolwide collection goal of 20,000 pounds within the next 3 weeks, urged that each student bring some recyclable paper to school every day and aim for individual goals of 120 pounds for the next 3-week period,[4] and explained that the money collected during this period would be used to purchase small trees for those students who requested them.

5. *Baseline 3* (approximately 3 weeks). Same as Baseline 2 except that on the 1st day, small trees were given to students upon request.

6. *Self-Recording* (approximately 2 weeks). A chart for recording individual participation was distributed to each of the 11 classrooms, and every class roster included an announcement of the group contingency that an extra 10 minutes of recess would be awarded the class if at least 75% of the students delivered at least one newspaper on 7 of the next 9 days.

7. *Baseline 4* (approximately 1 week). Same as Baseline 2 except that on the first day, the principal announced which classes had met the criterion for extra recess.

---

[3]It has been estimated that 120 pounds of recyclable paper save one tree from being cut down (NARI, 1974).

[4]The individual goal amounted to an average of 8 pounds of paper per student per day and, therefore, was probably unrealistic.

The large amount of paper collected by this recycling program by Hamad et al. (i.e., 19,749 pounds in 8 months) demonstrates the cost-efficiency of using public schools (and "piggyback" deliveries) for the recycling of resources used in the home (e.g., newspapers, aluminum, and glass containers). However, the comparisons between experimental conditions resulted in some unexpected and equivocal findings. Specifically, the feedback condition resulted in a mean of 157 pounds of paper per collection day, which was a decrease of approximately 50% from the mean across all baseline periods of 235 pounds. Increases above baseline were shown for the goal-setting and self-recording conditions (i.e., mean pounds per collection day of 366 and 296, respectively). Participation was markedly greatest during self-recording (i.e., 30% of the students participated compared to a mean of 2% participation for all other conditions), although only one class met the criterion for extra recess. Several factors could have contributed to the relatively low effects of the author's interventions. For example, the decline in paper deliveries from Baseline 1 to feedback may have been due to a decline in student, parent, and/or teacher interest following the initial prompt (i.e., memo) from the principal. A dissipation of the novelty of having a school paper drive may also have contributed to a decline in participation. Of course, it is also possible (even probable) that the students did not attend to the feedback thermometers, because there was no reinforcement contingency in effect. Further, the goal-setting intervention specified a somewhat unrealistic individual goal (i.e., 8 pounds of delivered paper per student per day) and added a reward component (i.e., a small tree upon request) that was not contingent upon recycling-related behaviors. Finally, a critical weakness of the self-recording intervention was the absence of a system for monitoring the students' daily recording of their participation in the recycling program.

The cost-benefit analysis by Hamad et al. (1979) indicated a net gain of \$247.68. The total cost of the program was estimated at \$177.50, which included \$41.20 for transporting the paper for resale, \$25 for the trees distributed during the goal-setting condition, and \$111.30 for labor. The benefit of \$425.18 was the sum of \$205.18 that was received from the sale of the newspapers, and the estimate of \$220 saved by preventing the newspapers from entering the solid waste stream. Another potential benefit (not included in the author's monetary calculations) is the reduced energy expenditure when recycled paper rather than wood pulp is used to manufacture paper. Indeed, such a savings has been estimated at 13 million BTUs of energy per ton of recycled paper (Hunt & Franklin, 1972). Furthermore, an extremely valuable (yet intangible) asset of school recycling programs is the impact on subsequent attitudes and behaviors of the participants. Indeed, personal involvement in a recycling program should not only increase awareness of conservation measures but should also increase the probability of future conservation action.

The paper-recycling program by Humphrey et al. (in press) demonstrated an approach to paper recycling that may be the most cost-effective wastepaper collection system that an average community could implement. The basic approach is cost-effective on two counts. It is workable at large institutions such as schools, hospitals, industrial plants, and government buildings, where large numbers of individuals may be readily organized for program participation and where the sources for recyclable wastepaper are plentiful; and the target is to recover high-grade rather than mixed paper. Compared to that for mixed paper, the monetary return for high-grade paper is much greater (i.e., from $50 to $100 per ton in 1976), and the market is more stable ("Paper Profits," 1977).

The 10-week wastepaper recovery program studied by Humphrey et al. essentially involved workers in administrative offices separating high-grade wastepaper as it was generated. The efficacy of three relatively effortless methods of separating office wastepaper were compared: (1) using two personal wastebaskets—one for salvageable paper and the second for other waste; (2) using a divided wastebasket with one section for recyclable paper and the other for nonrecyclable disposables; and (3) using one personal wastebasket for recyclable paper and a centrally located, public receptacle for nonrecyclable paper and other disposables.

The program was initiated in 16 experimental areas at Penn State University with a letter to all potential participants requesting cooperation in a waste separation project. For half of the experimental areas, the department head or office supervisor contacted each program participant on two occasions to inquire about problems with the system and to encourage continued participation. The dependent variable was the quality of the contents in the recycle containers as determined by weighing the recyclable and nonrecyclable material and then calculating the percent of recyclable paper per container.

Overall, 90% of the material discarded in the recycle container was salvageable wastepaper. The less convenient, central-can procedure resulted in 83.6% recyclable wastepaper in the office trash cans, which was a significantly poorer-quality return than either the two-wastebasket or divided-wastebasket procedures, which gave 92.3% and 92.7% recyclable paper, respectively. For each separation procedure, the weekly percentages indicated a rather consistent deterioration in the quality of wastepaper separation over time, with a marked increase in quality during Weeks 3 and 7 when the personal prompts were given in 50% of the experimental areas. However, the extent of the prompting effects cannot be realized, because the authors did not separate the weekly quality precentages for the prompt and no-prompt conditions. However, the authors did report a slightly higher overall separation quality for the group that received the personal prompts (i.e., 92.5% vs. 88.0%).

Unfortunately, Humphrey et al. did not indicate the pounds of salvageable paper collected nor the amount of money received from selling the paper, so that the cost-effectiveness and environmental impact of their program cannot be determined. However, the program does represent a practical model for a straightforward and economical procedure of tapping a copious source of high-quality wastepaper. The participants in this study did suggest that the program would be improved if a more detailed list of recyclable and nonrecyclable materials were distributed and if the trash can for recyclable materials were obtrusively marked. Given the steady decline in separation quality, other program refinements should include at least a more systematic procedure of prompting program participation, and should perhaps also incorporate an individual or group contingency to reinforce the quality and/or quantity of wastepaper separation responses.

An additional improvement in the paper separation program implemented by Humphrey et al. would be to make the separation process more convenient and contamination less convenient. The "desk-top container" system recommended by the EPA probably represents such an improvement ("Paper Profits," 1977). Each participant (e.g., office worker) is given a small (14-cubic-inch), desk-top container that holds a 1-to-13-week accumulation of high-grade, recyclable paper. When the container is full, the participant deposits the contents in a large, centrally located container. Such a system has been operating at EPA's headquarters in Washington, D.C., since November 1975 and currently involves about 2700 individuals recycling 12.5 tons of high-grade paper per month. During the first year of operation, EPA recovered 150 tons of high-grade paper and earned $11,000.

The incredibly low contamination rate for the desk-top container system (only 3%) is at least partially due to convenience. Not only is the special desk-top receptacle convenient for depositing used, high-grade office paper; it is also inconvenient (or sometimes impossible) to use for nonrecyclable paper and other disposables. The company that has manufactured and promoted the desk-top container system since the summer of 1972 (i.e., Shade Information Systems, Inc.) currently handles more than 400 accounts, with a variety of private businesses, universities, and government agencies; these accounts return over 750 tons of paper each month. Included among them is the National Bureau of Standards, which boasts a participation rate of 95% (out of 1400 employees) and a recovery rate of 32% that saves $2400 in disposable costs each year ("Paper Profits," 1977).

Arbuthnot, Tedeschi, Wayner, Turner, Kressel, and Rush (1977) observed that prompting might be effective in encouraging relatively inconvenient participation in a paper-recycling program *if* the intervention procedure also included effective prompting of more convenient recycling-related behaviors. Specifically, these investigators administered the following three antecedent conditions to individuals in Athens, Ohio, who were not participating in the community recycling program: (1) a door-to-door *survey,* whereby graduate students interviewed residents regarding their knowledge of recycling issues; (2) an *appeal* condition immediately following the survey, whereby the residents were given a handout of helpful recycling hints and a plastic bag in which to save recyclable cans for 1 week; and (3) a *letter* condition, whereby 2 weeks after the appeal condition, individuals were sent a letter that described the benefits of a communitywide recycling program and that included a request that the recipients mail an enclosed postcard to their city council representative in order to indicate their attitudes (positive or negative) toward expanding recycling programs in their community. Residents in a "three-step" group received each of these antecedent conditions; residents in "two-step" groups received one of the three possible pairs of conditions (i.e., survey and appeal, survey and letter, or appeal and letter); residents in a "one-step" group were given only one of the antecedent conditions; and a control group received none of the antecedent interventions.

The dependent variable in the Arbuthnot et al. study was the percentages of compliant subjects in each experimental condition who reported participating in the community recycling program 1 to 2 months and 18 months after the prompting intervention(s). Unfortunately, program participation was assessed via a "disguised telephone survey" and not with observations of actual recycling-related behaviors. The results showed that the probability of an individual reporting participation in the Athens recycling program was directly related to the number of prior prompts received *and* followed. For example, 87.5% of the 28 individuals who complied with the survey, appeal, and letter prompts reported participation in the recycling program 1 to 2 months after treatments. The reported participation rates were as follows for the compliant subjects in the other conditions: 80.0% of 23 individuals for the appeal-and-letter condition; 32.0% of 25 for survey and appeal; 20.0% of 28 for survey and letter; 8.3% of 24 for appeal alone; 0.0% of 24 for letter alone; 5.9% of 34 for survey alone; and 0.0% of 22 for the control group. Practically the same participation percentages were reported 18 months after treatment.

The authors reported that 19 individuals refused to respond to the survey (i.e., 14%) but did not report a participation rate for the other conditions. It would have been valuable to know what percentage of the individuals contacted (and prompted) in each condition saved cans in the distributed bags and/or wrote a town council member about recycling. Apparently, the authors overlooked such fine analyses of the participation rates per prompting procedure because their study was really not designed as a behavioral analysis of participation in a community recycling project, but rather was an attempt to demonstrate a field application of the Freedman and Fraser (1966) "foot-in-the-door" principle (i.e., that compliance with an initial small request will increase the probability of compliance with a subsequent, more effortful request). However, the authors' conclusion that compliance with three recycling-related requests produced participation in the community recycling program is not necessarily valid. Instead, it is reasonable to assume that those individuals who were most likely to comply with a prompt to participate in a recycling program were also those individuals who were most apt to comply with the prior request for recycling-related activity. In this regard, the multiple prompting technique was successful in selecting those individuals who would most likely emit relatively inconvenient behaviors in support of a community recycling project.

The thesis and dissertation of Harvey E. Jacobs at Florida State University (under the supervision of Jon S. Bailey) represents the most comprehensive behavioral analysis of residential resource recycling. For his master's thesis (Jacobs, 1978), Jacobs studied strategies for facilitating newspaper recycling among 700 residences in Tallahassee, Florida. The observation "sessions" were the first and third Saturdays each month; they consisted of driving by each of the households and picking up any newspapers or cans placed outside. The dependent variable was the percentage of homes participating per session for each of the five experimental groups defined next. Each group consisted of from 86 to 147 households.

1. *Information Only*. A flyer was delivered to the household every other week for 8 weeks, notifying the residents of the scheduled pickup and specified participation procedures.
2. *Penny-a-Pound*. The information flyer included a specification of the contingency that on the day following the collection, the participant would be paid a penny for each pound of newspaper.
3. *Lottery*. The information flyer included the definition of a variable-ratio contingency of reinforcement, whereby each participant became eligible for a $5 prize that was drawn randomly from each session's participating households and was distributed on the day following the collection.
4. *Weekly Pickup*. The information flyer specified a weekly collection of recyclable newspapers and cans, rather than the less frequent, biweekly collection for the other groups.
5. *Control*. Data were recorded from one group of residences that received no special information pamphlet nor any other experimental manipulation.

Each experimental group consisted of from 86 to 147 residences, and each experimental intervention was preceded by four baseline sessions (i.e., 8 weeks). Furthermore, 10 months prior to the start of the study, all the households had been notified of a biweekly collection service for recyclable paper and cans.

The percentage of households leaving recyclable materials for collection was quite low for each observation session and experimental group; but relative to the controls, each treatment intervention influenced a significant increase in participation. More specifically, the average percent participation during baseline was 2.84% for information only, 4.08% for penny-a-pound, 2.9% for lottery, 3.39% for weekly pickup, and 1.13% for the control; whereas during treatment, mean participation increased to 8.33% for information only, 9.04% for penny-a-pound, 14.24% for lottery, 8.75% for weekly pickup, and 2.04% for control. Although these changes indicate percentage increases of 193%, 120%, 390%, and 193% for the information only, penny-a-pound, lottery, and weekly pickup, respectively, the mean amount of recyclable newspapers collected per group did not change reliably across any observation session. Indeed, a comprehensive cost–benefit analysis indicated that no group generated enough revenue to cover the expense of operating the intervention strategy. Actually, the interventions provided for only about 25% to 33% of the revenues necessary to maintain the program. However, it is noteworthy that treatment lasted only 10 weeks (i.e., 5 observation sessions); more favorable cost–benefit ratios may result from a longer-term program.

For his dissertation research (as reported by Jacobs & Bailey, 1979a, 1979b), Dr. Jacobs accomplished a systematic behavioral analysis of a variety of potential determinants of residential recycling. The first experiment examined socioeconomic level as a correlate of participation in a community recycling program. Four neighborhoods were selected on the basis of their socioeconomic representation of lower, lower-middle, middle, and upper-middle income levels within Tallahassee–Leon County, Florida. The residents were informed of weekly curbside pickup of recyclable newspapers and cans by a brochure delivered to their door. After 4 to 6 weeks, a "reminder flyer" was distributed to the homes. For the 11-week study, participation was a direct function of socioeconomic level, with approximate mean participation percentages of 3%, 8%, 16%, and 25% for lower, lower-middle, middle, and upper-middle levels, respectively. The follow-up reminder flyer had no effect on the lower- and lower-middle-income residents, but did temporarily increase participation in the upper-middle- and middle-income neighborhoods (i.e., an initial increase of 10 to 12 percentage points).

In a second experiment, Jacobs and Bailey studied the advantage of having the collection service for recyclables scheduled on the same day as the garbage collection. For 11 weeks, the collection of recyclable cans and newspapers occurred on Tuesdays for a large neighborhood in which half the homes had their garbage collected on Tuesdays and Fridays and half had garbage collection service on Mondays and Thursdays. Each week, more residents with both garbage and recycling service on the same day rather than different days participated in the recycling program. The mean participation rate was 12.7% for the same-day-collection group and 7.5% for the different-day-collection group.

For a third field study, Jacobs and Bailey compared the effects of "door-to-door pamphleting" and "phoning campaigns" in encouraging residential recycling. A reminder flyer was delivered to the door of one group of residents in a large middle-income area 5 weeks after the initial announcement of a neighborhood recycling program; another group received the reminder handbill 11 weeks after program initiation. Phone reminders were given to two other groups in this neighborhood, 5 or 11 weeks after the start of the recycling program. Neither prompting procedure seemed to have consistent facilitative effects on program participation. Participation increased somewhat after the door prompt, but this change was temporary. The phone prompt appeared to have no effect on participation in the recycling program.

A fourth recycling study by Jacobs and Bailey compared two methods of notifying residents of a recycling program—door-to-door distribution of brochures versus a newspaper ad. Four days prior to the first scheduled pickup of recyclable cans and newspaper, a large, ⅛-page ad was run in the only local paper, announcing the recycling program for a particular subdivision of approximately 140 homes. Half the homes in the subdivision received an information brochure on the same day the newspaper ad appeared; the remaining homes received this brochure prior to the 9th observation session (i.e., 18 weeks after the newspaper ad).

The results of this study demonstrated marked advantages of the door-to-door brochure strategy over the newspaper ad. More specifically, the group that received only the newspaper ad at the start of the study showed consistently less participation during the first 8 collection periods than the group that received the newspaper ad and the brochure (mean participation during this period was 2.9% versus 7.6% for these groups). Furthermore, when the group that started with only a newspaper ad was given a brochure, participation more than doubled immediately. Mean participation for the 7 sessions following the brochure was about 7.9%, compared to a mean participation of 2.9% prior to the brochure intervention.

Three additional field experiments by Jacobs and Bailey investigated the beneficial effects of a special three-compartment, source separation container called a "Recyclit," which is sold by Sears, Roebuck and Co. For the first of these studies, residences in an upper-income area were divided into two groups of 63 houses each—a container group and a control group. An information brochure and the Recyclit were delivered to the container group, along with instructions and a number to call in case of questions; the controls received only the information brochure. Prior to the 10th collection week, a reminder brochure was delivered to all 126 residences. The results suggested marked advantages for the Recyclit container. More specifically, on all but 1 week of this 20-week study, the residents with a Recyclit showed prominently greater participation. The mean participation percentages were 19.3 and 7.8 for the container and control groups, respectively. The reminder brochure did not influence participation for either group.

The second experiment with the Recyclit containers was conducted in a lower-middle-income neighborhood. Four groups of 18 to 20 residences each were defined as follows. Twelve weeks after the initiation of the recycling program, each home in one group (Group C-1) received a Recyclit and instructions for its use. Each household in Group P-1 received a reminder-prompt with recycling instructions on the same day that Recyclits were delivered to C-1. Five weeks later, an additional group received the Recyclit (C-2), and another group received the door-delivered brochure (P-2). Seven weeks after these interventions, C-1 and P-2 became the targets for an intensive prompting campaign, receiving a different recycle-prompting handbill twice a week for 5 weeks.

The distribution of the Recyclits (Group C-1 and C-2) resulted in an immediate participation increase of about 10 percentage points, which declined after 4 weeks to near baseline levels of approximately 10% to 12% participation. The door-to-door distribution of reminder flyers had no apparent influence on recycling by Groups P-1 and P-2. The approximate baseline levels of 9% and 13% for Groups P-1 and P-2, respectively, remained intact after prompting. The twice-a-week prompting program increased participation by an average of about 10 percentage points for both container groups (C-1 and C-2) but had no influence on participation for the prompt groups (P-1 and P-2).

For a final experiment in this series, Jacobs and Bailey implemented an intervention package that included a variety of factors found previously to promote participation in a community recycling program. In particular, one of two experimental groups from a middle- to upper-middle-income area received the following treatment package after 7 weeks of baseline: (1) a Recyclit container with usage instructions; (2) a flyer reminding the resident of the weekly collection day and how to recycle; (3) a letter stating the contingency that a "We Recycle" sticker for mailbox application would be awarded a household that participated in the program at least once; and (4) twice-a-week delivery of different reminder-prompts for a consecutive 5-week period. The second group received no treatment strategy.

During baseline, the treatment and control groups displayed equivalent recycling participation. However, upon implementation of the intervention package, the treatment group demonstrated a participation increase of 25 percentage points. Indeed, participation for this group averaged approximately 50% and did not drop for 13 weeks, even after the termination of twice-a-week prompting.[5]

CONCLUDING COMMENTS

This chapter was written to identify issues regarding waste reduction and resource recovery for energy conservation, and to describe critically the research contributions of behavioral scientists that could be applied for the benefit of community resource conservation programs. Studies of correlations between attitudes and recycling-related behavior have identified some factors that may help to identify those people who are apt to participate in a resource recovery program (e.g., individuals who are relatively liberal, well educated, and of high socioeconomic status). Research that applied behavioral analysis to problems of

[5]At the time of this writing, data were still being collected for this study.

waste reduction and recycling has identified environmental factors that can increase the cost-effectiveness of a community recycling program (e.g., reinforcement contingencies, specially designed recycling containers, response-feedback techniques). Indeed, most of the behavioral studies have demonstrated that a cost-effective recycling program requires some sort of incentive program to encourage participation. However, it is this author's opinion that none of these studies were long-term nor large-scale enough to prescribe optimal conditions and contingencies for a communitywide resource conservation program.

It would certainly be beneficial if behavioral analysts would study the environmental determinants of recycling-related behaviors for longer periods of time and on a wider scale than prior research. However, it is this author's opinion that a more pressing problem is the large-scale dissemination of the information gathered from the behavioral analyses of recycling programs. Strategies for encouraging recycling and for evaluating the behavioral impact and cost-effectiveness of recycling programs have been shared at psychology conventions and published in psychology journals but are rarely disseminated to individuals developing policy for actual waste reduction and recycling. In fact, outside of our discipline, the utility of behavioral analysis for resource conservation is hardly realized. For example, the fact that a comprehensive, resource recovery plan that was recently applied citywide in Seattle, Washington, and Portland, Oregon, was based on the premise that people will participate in a recycling program if they are given specific instructions and a reasonable rationale (Duncan, 1975) suggests an unawareness of the behavioral evidence. And the noted absence of social scientists from the variety of interdisciplinary meetings on waste reduction and resource recovery that have occurred nationwide in recent years (e.g., Office of Solid Waste Management Programs, 1975; Seldman, 1977; Wahl & Bancroft, 1975) testifies to a lack of appreciation for the potential of behavioral science as a contributor to solving waste management problems. Furthermore, a recent solid waste bill has established teams of specialists in technical, marketing, financial, and institutional areas to provide nationwide assistance for state and local governments regarding resource conservation and recovery; but so far, no representative from the behavioral or social sciences has been included on these teams, which are referred to as "Resource Conservation and Recovery Panels" ("Solid Waste Bill," 1976).

Thus a challenge more urgent perhaps than refining and expanding the research designed to study ways of increasing and/or maintaining participation in waste reduction and recycling programs is to develop, evaluate, and refine strategies for disseminating the expertise of behavioral science for actual solutions of waste management problems. Indeed, behavioral science should truly become *applied* behavioral science only when principles or technologies developed in laboratory or field experiments are used effectively in the real world.

REFERENCES

Arbuthnot, J. The roles of attitudinal and personality variables in the prediction of environmental behavior and knowledge. *Environment & Behavior,* in press.

Arbuthnot, J., Tedeschi, R., Wayner, M., Turner, J., Kressel, S., & Rush, R. The induction of sustained recycling behavior through the foot-in-the-door technique. *Journal of Environmental Systems,* 1977, *6,* 353–366.

Bickman, L. Environmental attitudes and actions. *The Journal of Social Psychology,* 1972, *87,* 323–324.

Bree, W. R. *Antilittering and recycling research.* Symposium presentation at the meeting of the American Psychological Association, San Francisco, August 1977.

Buel, M. H. Local trashmen agree to collect recyclables. *Willamette Week,* August 29, 1977, pp. 1; 3.

Burgess, R. L., Clark, R. N., & Hendee, J. C. An experimental analysis of anti-litter procedures. *Journal of Applied Behavior Analysis,* 1971, *4,* 71–75.

Chapman, C., & Risley, T. R. Anti-litter procedures in an urban high-density area. *Journal of Applied Behavior Analysis,* 1974, *7,* 377–384.

Clark, R. N., Burgess, R. L., & Hendee, J. C. The development of anti-litter behavior in a forest campground. *Journal of Applied Behavior Analysis,* 1972, *5,* 1–5.

Community Environmental Council, Inc. *Recycling: The state of the art.* Santa Barbara, Calif.: Author, 1978.

Couch, J. V., Garber, T., & Karpus, L. Response maintenance and paper recycling. *Journal of Environmental Systems,* 1979, *8,* 127–137.

Dernbach, J. Michigan's bottle bill gives new momentum to national campaign. *Environment Action Bulletin,* April 30, 1977, 3–5.

Duncan, R. C. The "ORE plan" for recycling household solid waste: An alternative garbage collection system. *Compost Science,* January–February 1975, 24–32.

Duncan, R. C. An economic garbage collection and recycling service. *Compost Science,* January–February 1976, 12–25. (a)

Duncan, R. C. The role of the ORE plan in developing a nationwide recycling network. *Compost Science,* Summer 1976, 25–32. (b)

Environmental Protection Agency. *Resource recovery and source reduction* (SW-122), second report to Congress. Washington, D.C.: Office of Solid Waste Management Programs, 1974.

Fowler, S. Irony department: Capitol Hill switching to deposit containers. *Environment Action Bulletin,* September 4, 1976, 7–8.

Franz, M. This town accepts the composting challenge. *Compost Science,* 1971, May–June *12,* No. 3.

Freedman, J. L., & Fraser, S. C. Compliance without pressure: The foot-in-the-door technique. *Journal of Personality and Social Psychology,* 1966, *4,* 195–202.

Geller, E. S., Chaffee, J. L., & Ingram, R. E. Promoting paper-recycling on a university campus. *Journal of Environmental Systems,* 1975, *5,* 39–57.

Geller, E. S., Farris, J. C., & Post, D. S. Promoting a consumer behavior for pollution control. *Journal of Applied Behavior Analysis,* 1973, *6,* 367–376.

Geller, E. S., Wylie, R. C., & Farris, J. C. An attempt at applying prompting and reinforcement toward pollution control. *Proceedings of the 79th Annual Convention of the American Psychological Association,* 1971, *6,* 701–702. (Summary)

Greene, A. K. Bring 'em back, repack and save. *Proceedings, 1975 Conference on Waste Reduction.* Washington, D.C.: U.S. Environmental Protection Agency, 1975, 105–112.

Hamad, C. D., Bettinger, R., Cooper, D., & Semb, G. *Using behavioral procedures to establish an elementary school paper-recycling program.* Unpublished manuscript, 1979. (Available from

George Semb, Department of Human Development, University of Kansas, Lawrence, Kansas 66045).

Hamad, C. D., Cooper, D., & Semb, G. Resource recovery: The use of a group contingency to increase paper recycling in an elementary school. *Journal of Applied Psychology,* 1977, *62,* 768–772.

Hayes, S. C., Johnson, V. W., & Cone, J. D. The marked item technique: A practical procedure for litter control. *Journal of Applied Behavior Analysis,* 1975, *8,* 381–386.

Heberlein, T. A. *Beliefs about sanctions, norm activation and violation of the anti-littering norm.* Paper presented at the meeting of the Rural Sociological Society, Montreal, Canada, 1974.

Hendee, J. C. No, to attitudes to evaluate environmental education. *The Journal of Environmental Education,* 1971, *3,* 65.

Humber, N. Waste reduction and resource recovery—there's room for both. *Waste Age,* November 1975.

Humphrey, C. R., Bord, R. J., Hammond, M. M., & Mann, S. Attitudes and conditions for cooperation in a paper recycling program. *Environment & Behavior,* in press.

Hunt, R. G., & Franklin, W. E. *Environmental effects of paper recycling.* Paper presented at the 73rd National Meeting of the American Institute of Chemical Engineers, Minneapolis, Minnesota, 1972. Reprinted by the Midwest Research Institute, Kansas City, Missouri.

Ingram, R. E., & Geller, E. S. A community-integrated, behavior modification approach to facilitating paper recycling. *JSAS Catalog of Selected Documents in Psychology,* 1975, *5,* 327. (Ms. No. 1097)

Institute for Local Self-Reliance. *Composting in the city* (Community Self-Reliance Series). Washington, D.C.: Author, 1975.

Jacobs, H. E. *An analysis of motivational effects in encouraging participation in a residential recycling program.* Unpublished master's thesis, Florida State University, 1978.

Jacobs, H. E., & Bailey, J. S. *The Leon County recycling program: The development of an empirically derived communitywide resource recovery program.* Symposium presentation at Association for Behavior Analysis meeting, Dearborn, Michigan, 1979. (a)

Jacobs, H. E., & Bailey, J. S. *A preliminary analysis of motivational factors in a residential recycling program.* Symposium presentation at Southeastern Psychological Association meeting, New Orleans, 1979. (b)

James, D. A food industry perspective. *Proceedings, 1975 Conference on Waste Reduction.* Washington, D.C.: U.S. Environmental Protection Agency, 1975, 32–35.

Kiefer, I. *Incentives for tire recycling and reuse.* Washington, D.C.: U.S. Environmental Protection Agency, 1974.

King, H. D. A more positive strategy. *Proceedings, 1975 Conference on Waste Reduction.* Washington, D.C.: U.S. Environmental Protection Agency, 1975, 129–135.

LaBreque, M. Garbage—refuse or resource? *Popular Science,* June 1977, pp. 95–98; 166.

League of Women Voters of the U.S. *Recycle: In search of new policies for resource recovery* (Pub. No. 132). Washington, D.C.: League of Women Voters Education Fund, 1972.

Levenson, H. Perception of environmental modifiability and involvement in antipollution activities. *The Journal of Psychology,* 1973, *84,* 237–239.

Levenson, H. Involvement in anti-pollution activities and perceived negative consequences from pollution. *Perceptual and Motor Skills,* 1974, *38,* 1105–1106.

Loube, M. *Beverage containers: The Vermont experience.* Washington, D.C.: U.S. Environmental Protection Agency, 1975.

Luyben, P. D., & Bailey, J. S. Newspaper recycling behavior: The effects of reinforcement versus proximity of containers. *Environment & Behavior,* in press.

Margulies, W. P. Steel and paper industries look at recycling as an answer to pollution. *Advertising Age,* 1970, *41,* 63.

NARI, Recycling resources: A guide to effective solid waste utilization. *National Association of Recycling Industries Publication,* 1974.

Nietzel, M. T., Winett, R. A., MacDonald, M. L., & Davidson, W. S. *Behavioral approaches to community psychology.* New York: Pergamon Press, 1977.

Office of Solid Waste Management Programs. *Proceedings, 1975 Conference on Waste Reduction.* Washington, D.C.; U.S. Environmental Protection Agency, 1975.

Paper profits. *Garbage Guide* (Solid Waste Project), Environmental Action Foundation, 1977, No. 9.

Peterson, J. R. Environmental protection and productivity. *Proceedings, 1975 Conference on Waste Reduction.* Washington, D.C.: U.S. Environmental Protection Agency, 1975, 41–44.

Quarles, J. R., Jr. The need for action grows. *Proceedings, 1975 Conference on Waste Reduction.* Washington, D.C.: U.S. Environmental Protection Agency, 1975, 21–26.

Reid, D. H., Luyben, P. L., Rawers, R. J., & Bailey, J. S. The effects of prompting and proximity of containers on newspaper recycling behavior. *Environment and Behavior,* 1978, *8,* 471–483.

Resource recovery. *Garbage Guide,* Environmental Action Foundation, Washington, D.C., *8,* 1977.

Robinson, S. N., & Frisch, M. H. *Social-environmental influences on littering in a post office.* Paper presented at the meeting of the Eastern Psychological Association, New York, April 1975.

Ruvin, H. Beverage container legislation: The referendum in Dade County. *Proceedings, 1975 Conference on Waste Reduction.* Washington, D.C.: U.S. Environmental Protection Agency, 1975, 124–128.

Savage, J. F., & Richmond, H. R. III. Oregon's bottle bill: A riproaring success. *OSPIRG Reports,* Portland, Oregon, 1974.

Seldman, N. *Garbage in America: Approaches to recycling.* Washington, D.C.: Institute for Local Self-Reliance (1717 18th Street, N.W.), 1975.

Seldman, N. Collection/recycling systems challenge resource recovery. *Environmental Action Bulletin,* October 1976. (a)

Seldman, N. *Speech for technical information project's citizens and waste workshop.* St. Paul, Minnesota, October 1976. (Available from Neil Seldman, Director of Waste Utilization, Institute for Local Self-Reliance, 1917 18th Street, N.W., Washington, D.C. 20009). (b)

Seldman, N. Waste utilization—the trouble with high technology. *Self-Reliance,* September 1976, 11–12. (c)

Seldman, N. *New directions in solid waste planning.* Washington, D.C.: Institute for Local Self-Reliance (1717 18th Street, N.W.), April 1977.

Serrin, W. The Michigan bottle law: Model for a federal ban? *New York Times,* June 1979, p. 3.

Skinner, J. H. Effects of reuse and recycling of beverage containers. *Proceedings, 1975 Conference on Waste Reduction.* Washington, D.C.: U.S. Environmental Protection Agency, 1975, 136–139.

Smith, F. L., Jr. Waste paper recycling: Review of recent market demand and supply. *Pulp & Paper,* September 1975.

Solid waste bill becomes law. *Memo* (Solid Waste Project), Environmental Action Foundation, December 1976, pp. 1–2.

Steininger, M., & Voegtlin, K. Attitudinal bases of recycling. *The Journal of Social Psychology,* 1976, *100,* 155–156.

Tietjen, C., & Hart, S. S. Compost for agricultural land? *Journal of the Sanitary Engineering Division,* 1969, *95,* 269–287.

Van Tine, D. G. For a total systems approach. *Proceedings, 1975 Conference on Waste Reduction.* Washington, D.C.: U.S. Environmental Protection Agency, 1975, 113–118.

Voters ban throwaways in Michigan, Maine. *Memo* (Solid Waste Project), Environmental Action Foundation, December 1976, p. 1.

Waggoner, D. The Oregon bottle bill—facts and fantasies. *Environment Action Bulletin,* September 1976, 2-3.

Wahl, D., & Allison, G. *Reduce: Targets, means and impacts of source reduction.* Washington, D.C.: League of Women Voters Education Fund, 1975.

Wahl, D., & Bancroft, R. L. Solid waste management today . . . bring about municipal change. *Nation's Cities,* August 1975, 17-32.

Walker, E. A. Industry's commitment to waste reduction. *Proceedings, 1975 Conference on Waste Reduction.* Washington, D.C.: U.S. Environmental Protection Agency, 1975, 119-123.

Weigel, R. H. Environmental concern: The development of a measure. *Environment & Behavior,* in press.

Weigel, R. H., & Newman, L. S. Increasing attitude-behavior correspondence by broadening the scope of the behavioral measure. *Journal of Personality and Social Psychology,* 1976, *33,* 793-802.

Weigel, R. H., Woolston, V. L., & Gendelman, D. S. *Psychological studies of pollution control: An annotated bibliography.* Unpublished manuscript, Amherst College, 1977. (Available from Russell H. Weigel, Department of Psychology, Amherst College, Amherst, Mass. 01002).

Wendt, K. A. Approaches to source reduction. *Proceedings, 1975 Conference on Waste Reduction.* Washington, D.C.: U.S. Environmental Protection Agency, 1975, 66-77.

Westerman, R. R. Waste management through product design: The case of automobile tires. *Proceedings, 1975 Conference on Waste Reduction.* Washington, D.C.: U.S. Environmental Protection Agency, 1975, 83-94.

Witmer, J. F., & Geller, E. S. Facilitating paper recycling: Effects of prompts, raffles, and contests. *Journal of Applied Behavior Analysis,* 1976, *9,* 315-322.

6 Perception and Acceptability of Risk from Energy Systems

Paul Slovic
Baruch Fischhoff
Sarah Lichtenstein
Decision Research
A Branch of Perceptronics, Inc.

Over the next few decades, the success of energy production policies will depend vitally on public attitudes. Gradually, it has become recognized that energy decisions cannot be determined by technical criteria alone. Social, psychological, and political issues are crucial and involve such questions as: ''What kinds of risks should be accepted in exchange for what kinds of benefits? With how much uncertainty of specific kinds does the public care to live? How does one weigh the substantial routine impact of some technologies (for example, burning coal) with the small chance of a big disaster associated with others. . . .?'' (Holdren, 1976, p. 22).

Despite the importance of these questions, we lack knowledge about the social and psychological factors (goals, values, criteria, and so on) that determine public responses to technological risks in general or risks from energy systems in particular. This is less because the problems in this area are difficult (which they are) than because very little time, effort, and research funding have been applied to them. One reason for this neglect is the fact that scientists and policymakers have been slow to recognize the importance of public attitudes and perceptions. Writing in the *American Scientist,* Alvin Weinberg (1976) observed:

> As I compare the issues we perceived during the infancy of nuclear energy with those that have emerged during its maturity, the public perception and acceptance of nuclear energy appears to be the question that we missed rather badly. . . . This issue has emerged as the most critical question concerning the future of nuclear energy. [p. 19].

Fortunately, a flurry of research activity in recent years has provided the beginnings of insight into the sociopsychological dynamics of societal risk taking. What follows is a brief review of that work. It starts with a discussion of the difficulties that human beings experience when attempting to estimate risks. With this as background, the remainder of the chapter discusses the perception and acceptance of risks from energy systems, with particular emphasis on response to nuclear power.

COPING INTELLECTUALLY WITH RISK

Evaluation of the risks from various energy systems requires, of experts and lay people alike, an appreciation of the probabilistic nature of the world and an ability to think intelligently about unlikely but consequential events. As Weinberg (1976) noted in the context of nuclear power, "We certainly accept on faith that our human intellect is capable of dealing with this new source of energy [p. 21]." Recently, however, the faith of many who study human judgment and decision processes has been shaken.

Consider, for example, probabilistic reasoning. Because of its importance to decision making, the question of how people assess the probabilities of uncertain events has been a focus of research interest. This research indicates that intelligent people systematically violate the principles of rational decision making when judging probabilities, making predictions, or otherwise attempting to cope with uncertainty. Frequently, these violations can be traced to the use of judgmental heuristics, mental strategies by which people reduce difficult judgments to simpler ones (Tversky & Kahneman, 1974). These heuristics are useful guides in some circumstances, but in others, they lead to large and persistent biases with serious implications for decision making.

Availability Bias

This is not the place to pursue a full discussion of heuristics and biases in probabilistic thinking. Extensive reviews are available in a number of articles (Slovic, Fischhoff, & Lichtenstein, 1977; Slovic, Kunreuther, & White, 1974; Tversky & Kahneman, 1974). However, one heuristic bears mention here because of its special relevance to energy decisions. This is the "availability heuristic," whereby an event is judged likely or frequent if it is easy to imagine or recall instances of it. Generally, instances of frequent events are easier to recall than instances of less frequent events, and likely occurrences are easier to imagine than unlikely ones. Thus availability is often an appropriate cue for judging frequency and probability. However, availability is also affected by numerous factors unrelated to likelihood. As a result, reliance on it may lead people to exaggerate the probabilities of events that are particularly recent, vivid, or emotionally salient.

The notion of availability is potentially one of the most important ideas for helping us understand the distortions that occur in our perceptions of risks. For example, in discussing floodplain residents, Kates (1962) writes:

> A major limitation to human ability to use improved flood hazard information is a basic reliance on experience. Men on flood plains appear to be very much prisoners of their experience. . . Recently experienced floods appear to set an upward bound to the size of loss with which managers believe they ought to be concerned [p. 140].

Kates attributes much of the difficulty in improving flood control to the "inability of individuals to conceptualize floods that have never occurred" (Kates, 1962, p. 92). He observes that in making forecasts of future flood potential, individuals "are strongly conditioned by their immediate past and limit their extrapolation to simplified constructs, seeing the future as a mirror of that past" (Kates, 1962, p. 88). In this regard, it is interesting that the purchase of earthquake insurance increases sharply after a quake and then decreases steadily as the memories become less vivid (Steinbrugge, McClure, & Snow, 1969).

Availability bias is illustrated by several recent studies in which college students and members of the League of Women Voters were asked to judge the frequency of various causes of death, such as smallpox, tornadoes, and heart disease (Lichtenstein, Slovic, Fischhoff, Layman, & Combs, 1978). In one study, they were first told the annual death toll for motor vehicle accidents in the United States (50,000) and then were asked to estimate the frequencies of 40 other causes of death. In another study, participants were given 2 causes of death and were asked to judge which of the 2 was more frequent.

Both studies showed people's judgments to be moderately accurate in a global sense; that is, people usually knew which were the most and least frequent lethal events. However, within this global picture, people made serious misjudgments, many of which seemed to reflect availability bias. For example, accidents were judged to cause as many deaths as diseases, whereas diseases actually take about 15 times as many lives. Homicides were incorrectly judged to be more frequent than diabetes and stomach cancer. Homicides were also judged to be about as frequent as stroke, although the latter actually claims about 11 times as many lives. Frequencies of death from botulism, tornadoes, and pregnancy (including childbirth and abortion) were also greatly overestimated.

Table 6.1 lists the lethal events whose frequencies were most poorly judged in our studies. In keeping with availability considerations, overestimated items were dramatic and sensational, whereas underestimated items tended to be unspectacular events that claim one victim at a time and are common in nonfatal form. A follow-up study showed that newspaper coverage of the various causes of death was biased in much the same way as were people's judgments (Combs & Slovic, 1979).

TABLE 6.1
Bias in Judged Frequency of Death

Most Overestimated	*Most Underestimated*
All accidents	Smallpox vaccination
Motor vehicle accident	Diabetes
Pregnancy, childbirth, and abortion	Stomach cancer
Tornado	Lightning
Flood	Stroke
Botulism	Tuberculosis
All cancer	Asthma
Fire and flames	Emphysema
Venomous bite or sting	
Homicide	

Overconfidence

A particularly pernicious aspect of heuristics is that people are typically very confident about judgments based on them. For example, in a follow-up to the study on causes of death, participants were asked to indicate the odds that they were correct in their judgments about which of two lethal events was more frequent (Fischhoff, Slovic, & Lichtenstein, 1977). Odds of 100 : 1 or greater were given often (25% of the time). However, about 1 out of every 8 answers associated with such extreme confidence was wrong (fewer than 1 in 100 would have been wrong if the odds had been appropriate). About 30% of the judges gave odds greater than 50 : 1 to the incorrect assertion that homicides were more frequent than suicides. The psychological basis for this unwarranted certainty seems to be people's insensitivity to the tenuousness of the assumptions upon which their judgments are based (in this case, the validity of the availability heuristic). Such overconfidence is dangerous. It indicates that we often do not realize how little we know and how much additional information we need about the various problems we face.

Overconfidence manifests itself in other ways as well. A typical task in estimating failure rates or other uncertain quantities is to set upper and lower bounds so that there is a 98% chance that the true value lies between them. Experiments with diverse groups of people making many different kinds of judgments have shown that rather than 2% of true values falling outside the 98% confidence bounds, 20% to 50% do so (Lichtenstein, Fischhoff, & Phillips, 1977). People think that they can estimate such values with much greater precision than is actually the case.

Unfortunately, experts seem as prone to overconfidence as lay people. Hynes and VanMarcke (1976) asked seven "internationally known" geotechnical engineers to predict the height of an embankment that would cause a clay foundation to fail and to specify confidence bounds around this estimate that were wide enough to have a 50% chance of enclosing the true failure height. None of the bounds specified by these experts actually did enclose the true height. The multimillion dollar *Reactor Safety Study* (U.S. NRC, 1975), in assessing the probability of a core melt in a nuclear reactor, used a procedure for setting confidence bounds that has been found in experiments to produce a high degree of overconfidence. Related problems led a recent review committee to conclude that the *Reactor Safety Study* greatly overestimated the precision with which it had assessed the probability of a core melt (U.S. NRC, 1978).

Another case in point is the 1976 collapse of the Teton Dam. The Committee on Government Operations has attributed this disaster to the unwarranted confidence of engineers who were absolutely certain they had solved the many serious problems that arose during construction (U.S. Government, 1976). In routine practice, failure probabilities are not even calculated for new dams, even though about 1 in 300 fails when the reservoir is first filled. Further anecdotal evidence of overconfidence may be found in many other technical risk assessments (Fischhoff, 1977). Some common ways in which experts may overlook or misjudge pathways to disaster include:

- Failure to consider the ways in which human errors can affect technological systems. Example: Due to inadequate training and control room design, operators at Three Mile Island repeatedly misdiagnosed the problems of the reactor and took inappropriate corrective actions. A minor incident thus became a major accident.
- Overconfidence in current scientific knowledge. Example: The failure to recognize the harmful effects of X rays until societal use had become widespread and largely uncontrolled.
- Failure to appreciate how a technological system functions as a whole. Example: The rupture of a liquid-natural-gas storage tank in Cleveland in 1944 resulted in 128 deaths, largely because no one had realized the need for a dike to contain spillage. The DC-10 failed in several early flights because none of its designers realized that decompression of the cargo compartment would destroy vital parts of the plane's control system.
- Slowness in detecting chronic, cumulative environmental effects. Example: Although accidents to coal miners have long been recognized as one cost of operating fossil-fueled plants, the effects of acid rains on ecosystems were slow to be discovered.
- Failure to anticipate human response to safety measures. Example: The partial protection offered by dams and levees gives people a false sense of security and promotes development of the floodplain. When a rare flood does exceed the capacity of the dam, the damage may be considerably greater than if the floodplain had been unprotected.

Desire for Certainty

Every technology is a gamble of sorts, and like other gambles, its attractiveness depends on the probability and size of its possible gains and losses. Both scientific experiments and casual observation show that people have difficulty thinking about and resolving the risk/benefit conflicts even in simple gambles. One way to reduce the anxiety generated by confronting uncertainty is to deny that uncertainty. The denial inherent in this anxiety-reducing search for certainty is thus an additional source of overconfidence. This type of denial is illustrated by people who, when faced with natural hazards, view their world as either perfectly safe or as predictable enough to preclude worry. Thus, some flood victims interviewed by Kates (1962) flatly denied that floods could ever recur in their areas. Some thought (incorrectly) that new dams and reservoirs in the area would contain all potential floods, and others attributed previous floods to freak combinations of circumstances, unlikely to recur. Denial, of course, has its limits. Many people feel that they cannot ignore the risks of nuclear power. For these people, the search for certainty is best satisfied by outlawing the risk.

Scientists and policymakers who point out the gambles involved in societal decisions are often resented for the anxiety they provoke. Borch (1968) noted how annoyed corporate managers get with consultants who give them the probabilities of possible events instead of telling them exactly what will happen. Just before a blue-ribbon panel of scientists reported that they were 95% certain that cyclamates do not cause cancer, Food and Drug Administration Commissioner Alexander Schmidt said: "I'm looking for a clean bill of health, not a wishy-washy, iffy answer on cyclamates" (*Eugene Register-Guard,* January 14, 1976). Senator Edmund Muskie has called for "one-armed" scientists who do not respond "on the one hand, the evidence is so, but on the other hand . . ." when asked about the health effects of pollutants (David, 1975). In 1976, when people all over the country were demanding to know whether the swine flu vaccine was safe, the nature of their demands indicated that they were really trying to determine whether or not it was *perfectly* safe.

Perseverance of Beliefs

The difficulties of facing life as a gamble contribute to the polarization of opinions about hazards. For example, some people view nuclear power as extraordinarily safe, whereas others view it as a catastrophe in the making. It would be comforting to believe that these divergent beliefs would converge toward one "appropriate" view as new evidence is presented. Unfortunately, this is not likely to be the case. A great deal of research suggests that people's beliefs change slowly and are extraordinarily persistent in the face of contradictory evidence (Ross, 1977). Once formed, initial impressions tend to structure and distort the way in which subsequent evidence is interpreted. New evidence appears reliable and informative if it is consistent with one's initial beliefs; contradictory evidence is dismissed as unreliable, erroneous, or unrepresentative. Ross (1977) concluded his review of this phenomenon as follows:

> Erroneous impressions, theories, or data processing strategies, therefore, may not be changed through mere exposure to samples of new evidence. It is not contended, of course, that new evidence can never produce change—only that new evidence will produce *less* change than would be demanded by any logical or rational information-processing model [p. 210].

The Fallibility of Judgment

Our examination of risk perception leads to the following conclusions:

- Cognitive limitations, coupled with the anxieties generated by facing life as a gamble, cause uncertainty to be denied, risks to be distorted, and statements of fact to be believed with unwarranted confidence.
- Risk judgments are influenced (and sometimes biased) by the imaginability and memorability of hazards. People may, therefore, not have valid judgments even for familiar risks.
- Disagreements about risk should not be expected to evaporate in the presence of "evidence." Definitive evidence, particularly about rare hazards, is difficult to obtain. Weaker information is likely to be interpreted in a way that reinforces existing beliefs.

Because it can be shown that even well-informed lay people have difficulty judging risks accurately, it is tempting to conclude that the public should be removed from the decision-making process. The political ramifications of such a transfer of power to a technical elite are obvious. Indeed, it seems doubtful that such a massive disenfranchisement is feasible in any democratic society.

Furthermore, this transfer of decision making would seem to be misguided. For one thing, we have no assurance that experts' judgments are immune to biases once they are forced to go beyond documented evidence and rely upon judgment. Although judgmental biases have most often been demonstrated with lay people, there is evidence that the cognitive functioning of experts is basically like that of everyone else.

In addition, in many if not most cases, effective management of societal risks requires the cooperation of a large body of lay people. These people must agree to do without some things and accept substitutes for others; they must vote sensibly on ballot measures and for legislators who will serve them as surrogate hazard managers; they must obey safety rules and use the legal system responsibly. Even if the experts were much better judges of risk than lay people, giving experts an exclusive franchise on decision making would involve substituting short-term efficiency for the long-term effort needed to create an informed citizenry.

FORECASTING PUBLIC RESPONSE TOWARD NUCLEAR ENERGY

People respond to the hazards they perceive. The basic research cited earlier, supplemented by studies of perceptions of specific energy systems, could enable public response to those systems to be forecasted. To date, most studies of perceptions and attitudes toward energy systems have focused on nuclear power. In this section, we discuss some conclusions based on this research.

The General Problem

Even before the accident at Three Mile Island, the nuclear industry was foundering on the shoals of adverse public opinion. A sizable and tenacious opposition movement had been responsible for costly delays in the licensing and construction of new power plants in the United States and for political turmoil in several European nations.

The errant reactor at Three Mile Island has stimulated a predictable immediate rise in antinuclear fervor. Any attempt to plan the role of nuclear power in the nation's energy future must consider the determinants of this opposition and anticipate its future course. One clue lies in recent research showing that the images of potential nuclear disasters that have been formed in the minds of the antinuclear public are remarkably different from the assessments put forth by most technical experts. We next describe these images and speculate on their origins, permanence, and implications.

Political Realities

Although questions of safety are preeminent in the nuclear debate, it is important to recognize that opposition to nuclear power is an organized political movement fueled by many other concerns besides safety (Bronfman & Mattingly, 1976; Otway, Maurer, & Thomas, 1978; Wilkes, Lovington, Horne, Pulaski, & Poole, 1978). Whereas some nuclear opponents are motivated primarily by fear of routine or catastrophic radiation releases, others join the movement because they are disenchanted with growth, centralization, corporate dominance, technology, or government. These latter individuals may argue about safety because they view the hazardousness of nuclear power as its "Achilles Heel." Although the discussion that follows is not directly concerned with this larger political context, it does highlight the special qualities of nuclear power that cause (or allow) political opposition to be focused on considerations of risk.

Basic Perceptions

Opponents of nuclear power tend to believe both that its benefits are quite low and its risks are unacceptably great (Fischhoff, Slovic, Lichtenstein, Read & Combs, 1978). On the benefit side, people do not see nuclear power as a vital link in meeting basic energy needs (Pokorny, 1977). Many believe it to be merely a supplement to other sources of energy that are themselves adequate (or could be made adequate by conservation).

On the risk side, nuclear power appears to evoke greater feelings of dread than almost any other technological activity (Fischhoff, Slovic, Lichtenstein, Read & Combs, 1978). Some have attributed this reaction to fear of radiation's invisible and irreversible contamination, threatening cancer and genetic damage. However, use of diagnostic X rays, another radiation technology that incurs similar risks, is not similarly dreaded. If anything, people underestimate its risks. The difference may lie in another characteristic of nuclear power—its association with nuclear weaponry. As a result of its violent origins, nuclear power is viewed by people as a technology whose risks are uncontrollable, lethal, and potentially catastrophic—characteristics that are not present in people's perceptions of X rays.

Research in which people are asked to describe their mental images of a nuclear accident and its consequences reveals the uncontrollable, lethal, and catastrophic qualities of perceived nuclear risks. There is a widely held expectation that a serious reactor accident is likely within one's lifetime and could result in hundreds of thousands, even millions, of deaths (Slovic, Fischhoff, & Lichtenstein, 1979). In addition, such an accident is expected to cause severe, irreparable environmental damage over a vast geographic area. These expectations contrast dramatically with the nuclear industry's official view that multiple safety systems will limit the damage in the extremely unlikely event of a major accident.

One inevitable consequence of this "perception gap" is uncertainty and distrust on the part of a public that suspects that the risks are incomparably greater than the experts' assessments (Kasper, 1979). The experts, in turn, question the rationality of the public and decry the "emotionalism" stymieing technological progress. Bitter and sometimes violent confrontations result.

Recognition of the perception gap has led many experts to claim that the public must be "educated" about the "real" risks from nuclear power. One public opinion analyst (Pokorny, 1977) put the matter as follows:

> The biggest problem hindering a sophisticated judgment on this question is basic lack of knowledge and facts. Within this current attitudinal milieu, scare stories, confusion, and irrationality often triumph. Only through careful education of facts and knowledge can the people know what the real choices are [p. 12].

Our own view is that attempts designed to reduce the perception gap face major obstacles. This pessimistic conclusion is based on two key aspects of the problem—one technical and one psychological.

The Technical Reality

The technical reality is that there are few "cut and dried facts" regarding the probabilities of serious reactor mishaps. The technology is so new and the probabilities in question are so small that risk estimates cannot be based on empirical observation. Instead, risk assessments must be derived from complex mathematical models, such as the fault trees and event trees used in the *Reactor Safety Study* (U.S. NRC, 1975) to assess the probability and consequences of a loss-of-coolant accident. Despite an appearance of objectivity, risk assessments are inherently subjective. Someone, relying on judgment, must structure the analysis to determine the ways that failure might occur, their relative importance, and their logical interconnections.

The difficulties of performing risk assessments have led many critics to question their validity (Bryan, 1974; Fischhoff, 1977; Primack, 1975). One major concern is that important initiating events or pathways to failure may be omitted, causing risks to be underestimated. If omissions are as common and difficult to detect as suggested earlier, then underestimation may be substantial. Another problem in assessing the reliability of reactor designs is the difficulty of taking proper account of "common-mode failures." To ensure greater safety, many technological systems are highly redundant. Should one crucial part fail, there are others designed to do the same job or to limit the resulting damage. Because the probability of each part failing is very small, the probability of all failing—thereby creating a major disaster—should be minuscule. This reasoning is valid only if the various components are independent (so whatever causes one part to fail will not automatically cause the others to fail). "Common-mode failure" occurs when the independence assumption does not hold. For example, because electrical cables controlling the redundant safety systems at a reactor in Browns Ferry, Alabama, were not spatially separated, all five emergency core-cooling systems were rendered inoperative by a single fire. Developing models that take proper account of such contingencies is a very difficult enterprise. The potential for omissions is complicated by the tendency of problems that are out of sight to be effectively out of mind, their omission unrecognized (Fischhoff, Slovic, & Lichtenstein, 1978).

One critic's (Holdren, 1976) skepticism regarding the defensibility of assessments of rare catastrophes summarizes the technical problem concisely:

> The expert community is divided about the conceivable realism of probability estimates in the range of one in ten thousand to one in one billion per reactor year. I am among those who believe it to be impossible *in principle* to support numbers as small as these with convincing theoretical arguments. . . . The reason I hold this

view is straightforward: nuclear power systems are so complex that the probability the safety analysis contains serious errors . . . is so big as to render meaningless the tiny computed probability of accident [p. 21].

The Psychological Reality

Public fears of nuclear power should not be viewed as irrational. In part, these fears are fed by the realization that the facts are in dispute and that experts have been wrong in the past, as when they irradiated enlarged tonsils or permitted people to witness A-bomb tests at close range. Furthermore, experts' errors seem often to be in the direction of underestimating risks. What one can criticize, perhaps, is the extent to which people's fundamental ways of thinking (such as reliance on the availability heuristic) lead them to a distorted view of such information as is presented. Certainly the risks from nuclear power would seem to be a prime candidate for availability bias because of the extensive media coverage they receive and their association with the vivid, imaginable dangers of nuclear war. In contrast, the chronic, unspectacular effects of pollution associated with other energy sources may attract too little attention.

One disturbing implication of the availability heuristic is that any discussion of low-probability hazards, regardless of its content, may increase their memorability and imaginability and hence increase their perceived risks. This possibility poses a major barrier to open discussions regarding nuclear safety. Consider an engineer arguing the safety of disposing of nuclear wastes in a salt bed by pointing out the improbability of the various ways radioactivity could be accidentally released. Rather than reassuring the audience, the presentation might lead them to think, "I didn't realize there were that many things that could go wrong."

The availability heuristic magnifies fears of nuclear power by blurring the distinction between what is remotely possible and what is probable. As one nuclear proponent lamented: "When laymen discuss what *might* happen, they sometimes don't even bother to include the 'might' " (B. Cohen, 1974, p. 36). Another analyst (J. Cohen, 1972) has elaborated a similar theme in the misinterpretation of "worst-case" scenarios:

> It often has made little difference how bizarre or improbable the assumption in such an analysis was, since one had only to show that some undesirable effect could occur at a probability level greater than zero. Opponents of a proposed operation could destroy it simply by exercising their imaginations to dream up a set of conditions which, although they might admittedly be extremely improbable, could lead to some undesirable results. With such attitudes prevalent, planning a given nuclear operation becomes . . . perilous [p. 55].

Whereas the preceding discussion helps to clarify the source of the perception gap between pronuclear experts and their lay opponents, it does not point unambiguously to one side or the other as having the most accurate appraisal of the overall risks from nuclear power. Although memorability and imaginability are capable of enhancing public fears, inability to imagine all the possible ways that a system could fail might produce a false sense of security among technical experts. As a result, the identification of judgmental difficulties does not, in itself, afford an external criterion for closing the perception gap. Insofar as the actual risks may never be known with great precision and much new information is given to alternative interpretations, the gap may be with us for a long time. Thus, Three Mile Island "proved" the possibility of a catastrophic meltdown to some, whereas to others, it demonstrated the reliability of the multiple containment systems.

A Nuclear Future?

Are the strong fears and determined opposition to nuclear power likely to persist? Will nuclear power ever gain widespread public acceptance? Although answers to these questions are by no means clear, public response to X rays provides some clues. The almost universal acceptance of X rays shows that a radiation technology can be tolerated once its use becomes familiar, its benefits clear, and its practitioners trusted.

Nerve gas provides an enlightening case study. Few human creations could be more dreaded or more potentially catastrophic than this deadly substance. When, in December of 1969, the army decided to transfer nerve gas from Okinawa to the Umatilla Army Depot in Hermiston, Oregon, citizens of Oregon were outraged—except those in Hermiston. Whereas public opinion around the state was more than 90% opposed, residents of Hermiston were 95% *in favor* of the transfer (*Eugene Register-Guard,* December 18, 1969, and January 11, 1970). Several factors seem to have been crucial to Hermiston's acceptance of nerve gas. For one, munitions and toxic chemicals had been stored safely there since 1941 (so the record was good, and the presence of the hazard was familiar). Second, there were recognized economic benefits to the community from continued storage at the depot of hazardous substances, in addition to the satisfaction of doing something patriotic for the country. Finally, the responsible agency, the U.S. Army, was respected and trusted.

These examples illustrate the slow path through which nuclear power might gain acceptance. It requires an incontrovertible long-term safety record, a responsible agency that is respected and trusted, and a clear appreciation of benefit. In the aftermath of Three Mile Island, this path appears not only slow but unnavigable. A quicker path to acceptance, and one that may provide the only hope for the industry, could be forged by a severe energy shortage. Society has shown itself willing to accept increased risks in exchange for increased benefits. Brownouts, blackouts, or rationing of electricity would likely enhance the perceived need for nuclear power and increase public tolerance of its risks. One example of such a reaction is the oil crisis of 1973–1974, which broke the resistance to offshore drilling, the Alaska pipeline, and shale oil development, all of which had previously been delayed because of environmental concerns. Such crisis-induced acceptance of nuclear power may, however, produce anxiety, stress, and conflict in a population forced to tolerate what it perceives as great risk because of its addiction to the benefits of electricity.

ACCEPTANCE OF NONNUCLEAR ENERGY SYSTEMS

The arguments already presented suggest that problems of public acceptance should be much less severe for nonnuclear sources of energy. Fossil fuels and hydroelectric systems are familiar and common and appear to be perceived as less risky than they actually are (although data are needed to test this speculation). Fossil, hydroelectric, and solar energy systems have their origins in antiquity and work via mechanisms (combustion, water force, and sunshine) that are familiar, natural, and well understood. Accidents and fatalities with these systems tend to involve relatively few individuals who are spatially and socially isolated from the rest of society. Furthermore, their effect is consummated in a fixed period of time, without the threat of lingering consequences. Observation of the recent failure of the Teton Dam in Idaho suggests that even such catastrophic hazards as the collapse of a hydroelectric dam will quickly be forgotten by those not directly involved, in contrast to the consequences of a reactor accident.

CONCLUSIONS

Management of nuclear power and alternative energy systems must be based upon an understanding of how people think about risks. Our aim here has not been to document public opposition and fear of nuclear power, which are already well known. Instead, we have attempted to point out that this reaction stems both from recognition of the unresolved technical issues in the risk assessment process and from the fundamental thought processes that determine perceptions of risk. Nuclear hazards are particularly memorable and imaginable, yet hardly amenable to empirical verification. Their special qualities blur the distinction between the possible and the probable and produce an immense gap between the views of most technical experts and a significant portion of the public. This gap must be acknowledged, and the difficulty of reducing it by educational programs or empirical demonstrations of safety must be recognized by planners and policymakers.

Facing this problem means addressing some hard questions. Does technology force us to make decisions that cannot be made well (or successfully) in a democratic society? What kind of political institutions are needed to preserve democratic freedoms and ensure public participation when decisions involve extreme technical complexity, catastrophic risk, and great uncertainty?

ACKNOWLEDGMENTS

Support for this research was provided by the National Science Foundation under Grant PRA-7911934 to Clark University under subcontract to Perceptronics, Inc. Any opinions, findings and conclusions, or recommendations expressed in this chapter are those of the authors and do not necessarily reflect the views of the National Science Foundation.

REFERENCES

Borch, K. *The economics of uncertainty*. Princeton, N.J.: Princeton University Press, 1968.

Bronfman, L. M., & Mattingly, T. J., Jr. Critical mass: Politics, technology, and the public interest. *Nuclear Safety*, 1976, *17*, 539–549.

Bryan, W. B. Testimony before the Subcommittee on State Energy Policy; Committee on Planning, Land Use, and Energy; California State Assembly, February 1, 1974.

Cohen, B. L. Perspectives on the nuclear debate. *Bulletin of the Atomic Scientists*, 1974, *30*, 25–39.

Cohen, J. J. A case for benefit-risk analysis. In H. J. Otway (Ed.), *Risk vs. benefit: Solution or dream* (Report LA-4860-MS). Los Alamos Scientific Laboratory, February 1972. (Available from the National Technical Information Service).

Combs, B., & Slovic, P. Causes of death: Biased newspaper coverage and biased judgments. *Journalism Quarterly*, 1979, *56*, 837–843, 849.

David, E. E. Editorial. *Science*, 1975, *189*, 891.

Fischhoff, B. Cost-benefit analysis and the art of motorcycle maintenance. *Policy Sciences*, 1977, *8*, 177-202.

Fischhoff, B., Slovic, P., & Lichtenstein, S. Knowing with certainty: The appropriateness of extreme confidence. *Journal of Experimental Psychology: Human Perception and Performance*, 1977, *3*, 552–564.

Fischhoff, B., Slovic, P., & Lichtenstein, S. Fault trees: Sensitivity of estimated failure probabilities to problem representation. *Journal of Experimental Psychology: Human Perception and Performance*, 1978, *4*, 330–344.

Fischhoff, B., Slovic, P., Lichtenstein, S., Read, S., & Combs, B. How safe is safe enough? A psychometric study of attitudes towards technological risks and benefits. *Policy Sciences*, 1978, *9*, 127-152.

Holdren, J. P. The nuclear controversy and the limitations of decision making by experts. *Bulletin of the Atomic Scientists*, 1976, *32*(3), 20–22.

Hynes, M., & VanMarcke, E. Reliability of embankment performance predictions. *Proceedings of the ASCE Engineering Mechanics Division Specialty Conference*. Waterloo, Ontario, Canada: University of Waterloo Press, 1976.

Kasper, R. G. Perceived risk: Implications for policy. In G. Goodman & W. D. Rowe (Eds.), *Energy risk management*. London: Academic Press, 1979, pp. 87–95.

Kates, R. W. *Hazard and choice perception in flood plain management* (Research Paper No. 78). Chicago: University of Chicago, Department of Geography, 1962.

Lichtenstein, S., Fischhoff, B., & Phillips, L. D. Calibration of probabilities: The state of the art. In H. Jungermann & G. deZeeuw (Eds.), *Decision making and change in human affairs*. Amsterdam: T. Reidel, 1977.

Lichtenstein, S., Slovic, P., Fischhoff, B., Layman, M., & Combs, B. Judged frequency of lethal events. *Journal of Experimental Psychology: Human Learning and Memory,* 1978, *4,* 551–578.

Otway, H. J., Maurer, D., & Thomas, K. Nuclear power: The question of public acceptance. *Futures,* 1978, *10,* 109–118.

Pokorny, G. *Energy development: Attitudes and beliefs at the regional/national levels.* Cambridge, Mass.: Cambridge Reports, 1977.

Primack, J. Nuclear reactor safety: An introduction to the issues. *Bulletin of the Atomic Scientists,* 1975, *31*(7), 15–17.

Ross, L. The intuitive psychologist and his shortcomings: Distortions in the attribution process. In L. Berkowitz (Ed.), *Advances in experimental social psychology.* New York: Academic Press, 1977.

Slovic, P., Fischhoff, B., & Lichtenstein, S. Behavioral decision theory. *Annual Review of Psychology,* 1977, *28,* 1–39.

Slovic, P., Fischhoff, B., & Lichtenstein, S. Images of disaster: Perception and acceptance of risks from nuclear power. In G. Goodman & W. D. Rowe (Eds.), *Energy risk management.* London: Academic Press, 1979, pp. 223–245.

Slovic, P., Kunreuther, H., & White, G. F. Decision processes, rationality, and adjustment to natural hazards. In G. F. White (Ed.), *Natural hazards: Local, national and global.* New York: Oxford University Press, 1974.

Steinbrugge, K. V., McClure, F. E., & Snow, A. J. *Studies in seismicity and earthquake damage statistics* (Report [Appendix A] COM-71-00053). Washington, D.C.: U.S. Department of Commerce, 1969.

Tversky, A., & Kahneman, D. Judgment under uncertainty: Heuristics and biases. *Science,* 1974, *185,* 1124–1131.

U.S. Government. *Teton dam disaster.* Washington, D.C.: Committee on Government Operations, 1976.

U.S. Nuclear Regulatory Commission. *Reactor safety study: An assessment of accident risks in U.S. commercial nuclear power plants* (WASH 1400 [NUREG-75/014]). Washington, D.C.: The Commission, October 1975.

U.S. Nuclear Regulatory Commission. *Risk assessment review group report to the U.S. Nuclear Regulatory Commission* (NUREG/CR-0400). Washington, D.C.: The Commission, September 1978.

Weinberg, A. M. The maturity and future of nuclear energy. *American Scientist,* 1976, *64,* 16–21.

Wilkes, J. M., Lovington, M., Horne, R., Pulaski, F., & Poole, R. *Formation of attitudes about nuclear power.* Paper presented at the Third Annual Meeting of the Society for the Social Study of Science, Bloomington, Indiana, 1978.

7 Acceptance of a Nuclear Power Plant: Applications of the Expectancy-Value Model

Eric Sundstrom[1]
John W. Lounsbury[1]
University of Tennessee, Knoxville
Robert C. DeVault
Elizabeth Peele
Oak Ridge National Laboratory

This chapter summarizes the results of a study in which the expectancy-value model was applied to attitudes toward a nuclear power plant. The shortage of energy in the U.S. and the search for new sources of energy have raised many controversial questions, especially about nuclear power. Public debate has focused on such issues as whether to produce nuclear power, how to construct and safeguard nuclear reactors, and where to site the nuclear plants. At present (1979), 71 nuclear power plants are operating in the U.S., with 92 under construction and another 30 planned. Although nuclear plants may help meet the country's energy needs, their most serious drawback concerns the threat of nuclear accidents. The widely publicized accident at Three Mile Island near Harrisburg, Pennsylvania, almost resulted in a ''meltdown'' of the reactor core. It led to the venting of radioactive gasses and the temporary evacuation of part of the surrounding area. The accident heightened public concern about an already prominent social issue.

Despite considerable work on methods for social-impact assessment and application of those methods in case studies (see Finsterbusch & Wolf, 1977), researchers have paid scant attention to psychological responses related to nuclear power. We know very little about the structure of attitudes about nuclear power plants.

[1]Formerly consultants to the Energy Division, Oak Ridge National Laboratory.

The present study was designed to investigate the cognitive processes underlying public acceptance of a specific nuclear power plant. Our first decision concerned the choice of a model of the cognitive structure of responses to such a complex issue as nuclear power. Many available models have been applied to other topics, such as racial prejudice, drug addiction, and work motivation. The models include cognitive dissonance, cognitive consistency, self-perception, social comparison, information integration, and expectancy-value theory. We chose the expectancy-value[2] model for several reasons. First, the model has undergone extensive development. Second, empirical support exists for the model in both laboratory and field settings (see Fishbein & Ajzen, 1975; Mitchell, 1974; Mitchell & Biglan, 1971; Wahba & House, 1974). Third, the model allows the examination of multiple effects of a stimulus object, expectations about the potential effects (i.e., beliefs or perceived likelihoods) and subjective values placed on the effects (i.e., subjective evaluations, values, or utilities). This flexibility is important in the study of attitudes concerning nuclear power, which are likely to embody perceptions of both favorable and unfavorable outcomes that vary in subjective likelihood of occurrence.

Another decision concerned a research setting. We studied the responses of people living in a small community near what will be one of the world's largest nuclear plants. It is under construction approximately 4 miles from the town of Hartsville, Tennessee (population, approximately 5300). At the time of our study, a license had been granted to the Tennessee Valley Authority by the U.S. Nuclear Regulatory Commission to construct the plant, and an initial task force was on site to begin construction. The peak construction force of nearly 5500 was not scheduled to be on site until 1980. Articles in the local newspaper and interviews with key community members suggested that having a large nuclear plant practically in their back yard was a salient and controversial issue for the residents of Hartsville.

Predictions from the Expectancy-Value Model

The fundamental proposition of the expectancy-value model holds simply that "a person's attitude toward any object is a function of his beliefs about the object and the implicit evaluative responses associated with those beliefs" (Fishbein & Ajzen, 1975, p. 29). In algebraic terms,

$$A_o = \sum_{i=1}^{n} E_i V_i$$

[2]*Expectancy-value,* a term used by Fishbein and Ajzen (1975), is used here synonymously with the term *expectancy-valence,* a term more prevalent in industrial-organizational psychology literature (e.g., Mitchell, 1974). The following terms are also used synonymously: (1) *expectancy* and *perceived likelihood* (of occurrence of an outcome of the plant); and (2) *value* or *subjective values* and *desirability* (of occurrence of an outcome of the plant).

where A = the attitude toward the object, E_i = the subjective probability that the object is related to attribute i, V_i the evaluation of attribute i, and n is the number of attributes.

In applying the model to attitudes toward the nuclear plant among Hartsville residents, A represents the individual's attitude toward the power plant. We compiled a list of potential effects of the plant from three sources: a review of the literature on social impacts of nuclear power plants, interviews with officials from TVA and Hartsville and interviews with experts on nuclear power from Oak Ridge National Laboratory. Perceived likelihood of occurrence of these potential effects and the desirability of the effects are represented by E and V in the model, respectively.

We predicted that the degree to which Hartsville residents favored construction of the nuclear plant would depend on the extent to which they perceived desirable effects as relatively likely (e.g., higher-paying jobs and cheaper electricity) and undesirable effects as relatively unlikely (e.g., radiation hazards and overcrowding in the schools). We expected to find citizens opposed to construction of the plant to the extent that they saw it as likely to bring undesirable effects and unlikely to bring desirable effects.

We tested the predictions of expectancy-value theory using three approaches. The first involved simple $\Sigma\ EV$ scores as suggested in the original model (Vroom, 1964). We predicted that the sum of the products of the expectancies and valences would be strongly related to acceptance of the nuclear plant.

We also developed an extension of the model that involved grouping the outcomes on the basis of perceived likelihood. In the traditional approach, EV scores are simply summed to form a composite measure for subsequent analysis. This reveals nothing about which outcomes are perceived as "going together". Nor does it reveal which groups of outcomes are most strongly associated with acceptance or rejection of the plant. Such information would permit a more refined interpretation of the reasons for acceptance of the plant. We proposed to factor-analyze the E scores (perceived likelihood of effects), then compute $\Sigma\ EV$ scores for each set of potential effects identified by the factor analysis. We predicted that these factor-based $\Sigma\ EV$ scores would be at least as strongly related to acceptance of the plant as the simple sum of EV scores.

In view of criticism about the use of multiplicative scores (Schmidt, 1973), we incorporated still another variant of the model. In an earlier study involving the measurement of only the perceived likelihood of effects (Sundstrom, Lounsbury, Schuller, Fowler, & Mattingly, 1977), we found acceptance of the nuclear plant directly related to E scores derived from a factor analysis of perceived likelihoods of effects. Research on work motivation and job effort (e.g., Motowidlo, Loehr, & Dunnette, 1972; Schuster, Clark, & Rogers, 1971) has found similar results. For the present study we planned to group the E scores using factor analysis. We predicted that the $\Sigma\ E$ scores for groups of effects would be significantly related to acceptance of the nuclear plant.

METHOD

In a survey of 288 residents of the Hartsville area in August 1975, we conducted personal interviews that included closed-ended questions using 7-step scales to measure the perceived likelihood and desirability of 27 potential effects of the nuclear plant. The survey included two items to measure acceptance of the facility. A separate validation sample of 97 residents was interviewed using the same procedure.

Participants

Participants in the survey were members of two samples. The *primary sample* consisted of 288 people who had been interviewed in January 1975 in a similar survey conducted among 350 residents of the Hartsville area (see Lounsbury, Sundstrom, Schuller, Mattingly, & DeVault, 1977; Schuller, Fowler, Mattingly, Sundstrom, Lounsbury, Passino, Dowell, & Hutton, 1975; Sundstrom, Lounsbury, Schuller, Fowler, & Mattingly, 1977). They are part of a longitudinal study of residents' views of the nuclear plant (see Sundstrom, Costomiris, DeVault, Dowell, Lounsbury, Mattingly, Passino, & Peele, 1977). The second sample consisted of 97 people selected especially for the survey in August 1975. They were included in the survey to assess the effects, if any, of the January survey on responses in August. For present purposes, they are treated as a validation sample.

Selection of the Primary Sample. For the survey in January 1975, we drew a simple random sample of residents of the county. To list the adult residents, we began with two published directories (see Sundstrom, Costomiris, DeVault, Dowell, Lounsbury, Mattingly, Passino, & Peele, 1977) and, with the help of the local residents, updated the list. It included about 3600 residents of Trousdale County over the age of 18 years, and we assumed it was nearly exhaustive. For the survey in January, we randomly chose 375 respondents and 75 alternates. We contacted 422 people to obtain a sample of 350 participants (21 were unable to arrange an appointment, and 51 refused to participate; these 72 people were replaced with alternates). For the survey in August, we attempted to contact all 350 of the original sample. A total of 288 people or 82% participated. (Of the January participants, 8 had moved out of the county, 22 refused to be reinterviewed, and we were unable to contact 32.)

Selection of the Validation Sample. From the same list used in January, we selected a stratified random sample of 122, including 46 blacks (to offset a shortage of blacks in the panel). A total of 97 completed the survey; 21 refused to be interviewed, and 4 could not be contacted.

Characteristics of the Participants. Most of the participants were raised in rural areas (71%) or small towns (22%). Approximately half had lived in the county for more than 30 years. Most were married (79%). About half were women (52%). The average respondent was 46.6 years of age and had completed 11.7 years of school. Most respondents were employed (71%), primarily as managers, proprietors, foremen, clerical workers, laborers, salespersons, and craftsworkers. The primary and validation samples were different in that the primary sample contained 10% blacks and the validation group contained 36% blacks. The validation group also contained more persons in blue-collar occupations (e.g., labor, service).

Procedures

In consultation with community leaders, we hired 18 local residents as interviewers. All but 1 were women. Before collection of the data, the interviewers were told the purposes of the study and trained in techniques of interviewing.

Respondents first received a letter describing the survey and emphasizing that participation was voluntary and that participants would be paid. An interviewer contacted each respondent a few days later to arrange an interview. The interviewers asked questions from a printed schedule and wrote the responses word for word on the interview form. Interviews lasted about an hour. A few weeks later, respondents received checks for $5 as payment for their participation.

Measures

This chapter concerns acceptance of the nuclear plant, expected likelihood of its potential effects, and subjective value of the effects for the respondents. (See Sundstrom, Costomiris, DeVault, Dowell, Lounsbury, Mattingly, Passino, & Peele, 1977, for results related to other aspects of the survey.)

Acceptance of the Nuclear Plant. In assessing acceptance of the nuclear plant, we asked two questions. The one we considered most important was: "If it were up to you, would you permit construction of the TVA power plant near Hartsville?" Respondents were asked to consider four alternative answers: "definitely yes," "probably yes," "probably no," and "definitely no." This "forced-choice" format allowed no explicit "unsure" response.

A second open-ended question asked respondents to indicate how close to their homes, in miles, they would permit the plant to be built. Responses were recorded verbatim and ranged from .25 mile to 25,000 miles.

As our measure of acceptance of the nuclear plant, we formed a composite of the two items. We placed responses to the "permit construction" question on a 4-step scale, where "definitely yes" was 1.0 and "definitely no" was 4.0. We then converted this score to a z score (or standardized score, which is the number of standard deviations from the mean). We scored responses to the "distance from home" item as percentile ranks (for each response, the percent of respondents who made an equivalent or lower response). Our intention was to reduce the distortion reflected in extremely large distances. We then converted the percentile rank to a z score. The composite acceptance score was the sum of the two z scores. (The correlation of each z score and the composite was .91 with $N = 261$.) The composite had a mean of 0.22 with a standard deviation of 2.06. Negative scores reflect greater acceptance (would permit, would allow close to home); positive scores reflect opposition (would not permit, would want far from home).

Perceived Expectancies and Subjective Values of Potential Effects. Interviewers gave respondents a list of 27 possible effects of the nuclear plant, such as "traffic congestion" and "more jobs" (see the list in Table 7.2). Respondents were first asked to estimate how likely each effect was, from a list of seven choices: "certain to happen," "very likely," "likely," "perhaps," "unlikely," "very unlikely," and "almost impossible." The *E*, or likelihood, scores ranged from −3 ("certain") to +3 ("almost impossible"). After rating the likelihood of each possible effect, respondents rated their desirability, again from seven choices: "extremely desirable," "very desirable," "desirable," "neutral," "undesirable," "very undesirable," and "extremely undesirable." We scored both items on a 7-step scale centered on 0. Subjective value, or desirability, scores ranged from −3 ("extremely desirable") to +3 ("extremely undesirable").

RESULTS

In describing the results, we first present the data from the primary sample of 288 respondents, including the distribution of acceptance scores, mean scores for expectancies and subjective values, the two approaches to the expectancy-value model, and the approach based on the expectancy scores alone. We then present the same results from the validation sample.

Acceptance of the Nuclear Plant in the Primary Sample

As shown in Table 7.1, approximately two-thirds of the primary sample said that if they could decide, they would permit construction of the nuclear plant. Responses to the "permit construction" question were highly correlated with the "distance from home" question ($r = .66$; $n = 261$; $p < .001$) and with responses to the same "permit construction" question 6 months earlier ($r = .76$; p

TABLE 7.1
Acceptance of the Hartsville Nuclear Plant in the Primary Sample ($N = 261$)

Responses to the "Permit Construction" Question[a]	Responses to the "Distance From Home" Question[b] Less than 4 miles (N = 40)	5 miles[c] (N = 79)	6-19 miles (N = 38)	20-100 miles (N = 37)	More than 100 miles[d] (N = 67)	Percent of All Responses (N = 261)
"Definitely yes" ($N = 99$)	30	53	15	1	0	38%
"Probably yes' ($N = 81$)	9	20	18	19	15	31%
"Probably no" ($N = 36$)	1	2	3	10	20	14%
"Definitely no" ($N = 45$)	0	4	2	7	32	17%
Percent of all responses[e] ($N = 261$)	15%	30%	15%	14%	26%	100%

[a]"If it were up to you, would you permit construction of the TVA power plant near Hartsville?"

[b]"If you could decide where the plant were to be located, how close to your home, in miles, would you permit it to be built?"

[c]The actual site is 5 miles from Hartsville. This category includes responses indicating that: (1) the distance does not matter ($N = 21$); (2) the present site is acceptable ($N = 8$); or (3) the present site is acceptable under a specified condition ($N = 14$).

[d]Included in this category are responses indicating "as far as possible" or "would not permit."

[e]Excludes people who did not answer one or both of the questions. For all respondents to the "permit construction" question ($N = 285$), the distribution of responses is as follows: "definitely yes"–38%; "probably yes"–31%; "probably no"–13%; "definitely no"–18%.

$< .001$; see Sundstrom, Costomiris, DeVault, Dowell, Lounsbury, Mattingly, Passino, & Peele, 1977).

Expectancies and Subjective Values of Potential Effects in the Primary Sample

Expectancies. As shown in Table 7.2, respondents' average estimates of the likelihood of specific effects of the nuclear plant ranged from 1.88 ("very likely") to −0.56 ("unlikely"). Effects generally seen as "very likely" included traffic congestion, crowded schools, increased noise, shortage of housing, increased business, and meeting of new people. Respondents generally seemed unwilling to rule out any of the effects. Mean estimates of likelihood never reached −1.0 ("unlikely"), and only three potential effects had means lower than zero ("perhaps"). These effects, seen as least likely, included radiation hazard, sabotage at the plant, and cheap electricity.

Grouping of the Potential Effects. To classify the potential effects into groups seen as going together, or as having correlated chances of occurrence, we performed a factor analysis. We employed a principal components solution with varimax rotation and no iterations to extract factors with eigenvalues greater than 1.0. (This procedure differs from a principal components analysis only in that the principal components were rotated once.) Shown in Table 7.2, five factors emerged: (1) disruptive effects of population growth, such as traffic congestion and crowded schools, which we called "social disruption"; (2) hazards to safety

TABLE 7.2
Perceived Likelihood of 27 Potential Effects in the Primary Sample, Grouped on the Basis of a Principal Components Factor Analysis

Group of Effects	*Effects*	*Likelihood Scores*[a] *Mean*	*S.D.*	*Factor Loading on Assigned Factor*	*Mean*
Factor I	Traffic Congestion	1.88	1.17	.69	1.5
Disruptive Effects	More taverns and bars	1.45	1.07	.68	"very
of Growth	Crowding in schools	1.63	1.10	.67	likely"
(23% of variance)	Increased noise	1.50	1.09	.68	
	Increased crime	1.20	1.17	.65	
	Drugs in schools	1.30	1.19	.56	
	Increased taxes	0.93	1.19	.56	
	Housing shortages	1.76	1.02	.50	
Factor II	Radiation hazard	-0.03	1.41	.81	0.1
Hazards to	Air pollution	0.53	1.28	.76	"perhaps"
Safety and	Sabotage at plant	-0.50	1.28	.72	
Environment	Foggy days	0.10	1.42	.70	
(19% of variance)	Pollution of lake	0.49	1.35	.64	
Factor III	More stores & shopping	1.19	1.05	.79	1.0
Increased	More entertainment	0.84	1.23	.62	"likely"
Business and	More recreation areas	0.88	1.16	.58	
New Facilities	More billboards	0.62	1.17	.53	
(6% of variance)	Increased business	1.53	0.94	.53	
Factor IV	Town/tourist attraction	0.63	1.36	.74	1.2
Outside	Recognition of town	1.49	1.15	.67	"likely"
Attention	Meeting new people	1.66	1.02	.59	
(4% of variance)	Industrial development	0.97	1.15	.62	
Factor V	Increased land values	1.44	1.19	.72	1.1
Economic	More jobs	1.43	1.05	.62	"likely"
Benefits of	Better-paying jobs	1.30	1.15	.53	
Growth	Better schools	0.34	1.41	.42	
(4% of variance)					
Other effects[b]	Cheap electricity	-0.56	1.47		

[a]"Certain" = +3.0; "very likely" = +2.0; "likely" = +1.0; "perhaps" = 0; "unlikely" = -1.0; "very unlikely" = -2.0; "almost impossible" = -3.0.

[b]Effects with all loadings under .50 and unclassifiable in any of the groups.

or environment, such as radiation; (3) increases in business and new facilities that might accompany new business, such as stores and billboards; (4) attention from outside town, including tourism and "new people"; and (5) economic benefits of growth, such as more and better-paying jobs.

Subjective Values of Potential Effects. As shown in Table 7.3, mean ratings of the desirability of effects ranged from −2.1 ("very undesirable") to nearly +2.0 ("very desirable"). Effects rated least desirable included "increased

TABLE 7.3
Perceived Desirability of 27 Potential Effects and Correlations in the Primary Sample between Likelihood and Desirability

Group of Effects	Effects	Desirability Scores[a] Mean	S. D.	Correlation of Likelihood and Desirability	Mean
Disruptive Effects of Growth	Traffic congestion	-1.42	1.15	-.20**	-1.8 "very undesirable"
	More taverns and bars	-1.35	1.41	-.25**	
	Crowding in schools	-1.87	1.00	-.38**	
	Increased noise	-1.62	1.15	-.13*	
	Increased crime	-2.12	1.05	-.14*	
	Housing shortages	-1.46	0.93	-.25**	
Hazards to Safety and Environment	Radiation hazard	-2.01	1.06	-.07	-1.7 "very undesirable"
	Air pollution	-1.66	1.04	-.18**	
	Sabotage at plant	-1.89	1.17	.07	
	Foggy days	-1.40	1.03	-.20**	
	Pollution of lake	-1.65	1.07	-.22**	
Increased Business and New Facilities	More stores & shopping	1.53	0.95	.17**	1.0 "desirable"
	More entertainment	1.28	1.03	.07	
	More recreation areas	1.47	0.97	.03	
	More billboards	-0.62	1.30	.24**	
	Increased business	1.60	0.97	.16**	
Outside Attention	Town/tourist attraction	0.39	1.27	.35**	0.9 "desirable"
	Recognition of town	0.94	1.03	.32**	
	Meet new people	1.21	0.72	.22**	
	Industrial development	1.17	0.95	.26**	
Economic Benefits of Growth	Increased land values	1.04	1.32	-.03	1.7 "very desirable"
	More jobs	1.72	0.95	.24**	
	Better-paying jobs	1.98	1.02	.10**	
	Better schools	1.95	1.09	.16**	
Other Effects	Cheap electricity	1.87	1.15	-.01	

[a]"Extremely desirable" = +3.0; "very desirable" = +2.0; "desirable" = +1.0; "neutral" = 0; "undesirable" = -1.0; "very undesirable" = -2.0; "estremely undesirable" = -3.0.

$^{*}p < .05$.
$^{**}p < .01$.

crime'' and ''drugs in schools''; both were rated as ''extremely undesirable.'' Twelve other outcomes were rated ''undesirable'' (−0.5 or lower), including all of the seven disruptive effects of population growth and all of the effects classified as ''hazards.'' Not surprisingly, the effects rated most desirable were in the group called economic benefits of growth. ''Better-paying jobs'' and ''more jobs'' were both rated as ''very desirable.'' Effects classed as increased business and new facilities, attention from outside the area, and economic benefits of growth all were rated as desirable.

Relationship of Expectancies and Subjective Values. Considering the groups of effects, there was just one exception to the tendency to rate desirable events as likely and undesirable events as relatively unlikely. The exception was the group of disruptive effects of growth, which was seen as undesirable but likely.

Considering the 27 effects separately, 20 of the 27 correlations between ratings of likelihood and desirability were significant ($p < .05$), although the correlations were modest. (This finding contradicts the implicit assumption of expectancy-value theory that the E and V scores are independent.) As shown in Table 7.3, however, the signs of the correlations differed among the groups of effects. For the two undesirable groups of effects, social disruption and hazards, ratings of desirability and likelihood were almost all *inversely* related. Relatively low ratings of desirability were associated with relatively high ratings of likelihood—that is, the more undesirable the effect, the more likely. However, this relation did not hold for two of the most undesirable effects, radiation hazard and sabotage at the plant.

For the three groups of effects rated as desirable, the relationship of desirability and likelihood was predominantly *positive*—that is, the more desirable the effect, the more likely. Apparently, extreme ratings of desirability on many of the effects were associated with relatively high ratings of likelihood. By implication, the multiplicative scores used in the valence model may tend to give most weight to the extremely desirable or undesirable effects.

Traditional Expectancy-Value Model

Using the formula that defines the expectancy-value model, we formed EV scores by multiplying the expectancy- and subjective-value ratings for each of the 27 effects, summing the products for each respondent, and dividing by the number of effects to preserve the scale from −9 to +9. In algebraic terms, the EV score was

$$\Sigma\, EV = \frac{1}{k} \sum_{i=1}^{k} E_i\ (\text{Expectancy}_i) V_i (\text{value}_i)$$

(For respondents who omitted ratings of any of the effects, K was adjusted accordingly.)

The correlation between the ΣEV scores and composite acceptance scores was $-.70$ ($n = 261$; $p < .001$). As predicted, acceptance was associated with the perception that desirable effects were likely and undesirable effects were unlikely. Low degrees of acceptance were associated with perceptions of undesirable effects as likely and desirable effects as unlikely. The traditional approach to the valence model accounted for nearly half of the variance (49%) in the acceptance scores in the primary sample, providing strong support for the valence model.

Factor-Score Approaches

Factor Scores Based on EV *Scores.* To form factor scores, we simply summed the *EV* scores for the effects in each of the five groups and divided by the number of effects in the group (this amounts to unit weighting). Our procedure yielded average *EV* for each group of effects, which only approximated the weighted factor scores we might have obtained through other procedures.[3] We used unit weighting because the more sophisticated approaches all involve transformations of the raw scores, which might yield stronger findings but might also prove less amenable to cross-validation (cf. Alwin, 1973; Susmilch & Johnson, 1975).

In a multiple linear regression analysis, we used the composite acceptance score as the criterion variable and the five average *EV* scores as predictors. As shown in Table 7.4, the predictors were far from independent. Average *EV* scores were highly correlated for the two groups of undesirable effects (social disruption and hazards); scores were also highly intercorrelated for the three groups of desirable economic effects. Table 7.5 shows the results of the regression analysis, in which four of the five predictors accounted for 51% of the variance in acceptance scores. (The predictors entered the equation in order of the increment in the multiple correlation coefficient—that is, in order of the size of correlation with the criterion.) The strongest predictors were *EV* scores for hazards and increased business; acceptance was greater if hazards were seen as unlikely and increased business was seen as likely. All of the predictors were significantly related to acceptance scores, but the outside-attention group showed the lowest correlation and failed to account for a significant portion of the variance in acceptance scores. These results agree with the hypothesis, with the factor-based *EV* approach accounting for slightly more variance than the traditional simple-sum *EV* approach to the expectancy-value model.

Factor Scores Based Only on E *Scores.* As a second method of forming factor scores, we summed the expectancy scores for each factor-based group of

---

[3]We could, for example, have calculated mathematically independent factor scores using factor-score coefficients for all 27 outcomes, or we could have estimated them by using the outcomes with the highest loadings on each factor. See Gorsuch (1974).

TABLE 7.4
Intercorrelations Among Average Likelihood x Desirability Scores in the Primary Sample for the Five Groups of Potential Effects ($N = 261$)

Group of Effects	Group of Effects: *I*	*II*	*III*	*IV*	*V*
I. Social Disruption	–				
II. Hazards	.50**	–			
III. Increased Business	.12	.29**	–		
IV. Outside Attention	.03	.21**	.64**	–	
V. Economic Benefits	.06	.28**	.66**	.60**	–

$**p < .01$.

effects and divided by the number of effects, yielding an average perceived likelihood for each group of effects. Table 7.6 presents the intercorrelations among *E* scores, which closely resembled the intercorrelations among *EV* scores. The multiple regression analysis of acceptance scores also yielded similar results, as shown in Table 7.7. As before, all of the predictors were correlated with acceptance scores, but the correlations for outside attention and increased business were substantially lower. The analysis accounted for 51% of the variance in acceptance scores with only three predictors: hazards, economic benefits, and social disruption. The group of effects labeled "increased business" failed to account for a significant increment in variance when we used only the likelihood scores, but it did contribute significantly when we used the *EV* scores.

Results for the Validation Sample

Acceptance of the Nuclear Plant. In the validation sample of 97 people, 88 answered the "permit construction" question. (The 9 people who declined to

TABLE 7.5
Multiple Regression Analysis of Acceptance Scores in the Primary Sample Using Likelihood x Desirability for Five Groups of Effects as Predictors ($N = 261$)

Predictor	*Simple r*	*Multiple r*	*Multiple r^2*	*Change in r^2*	*F value (df = 1,255)*
Hazards	-.66**	.61	.38	.39	1.62**
Increased business	-.42**	.67	.45	.06	5.82*
Social disruption	-.50**	.71	.50	.05	31.34**
Economic benefits	-.39**	.72	.51	.01	3.85*
Outside attention[a]	-.34**	(.72)	(.51)	(0)	–

[a]Not in the regression equation.
$*p < .05$.
$**p < .01$.

TABLE 7.6
Intercorrelations Among Average Likelihood Scores in the Primary Sample for the Five Groups of Potential Effects ($N = 261$)

Group of Effects	*I*	*II*	*III*	*IV*	*V*
I. Social Disruption	–				
II. Hazards	.56**	–			
III. Increased Business	.09	-.10	–		
IV. Outside Attention	.14	-.08	.55**	–	
V. Economic Benefits	-.08	-.26**	.52**	.57**	–

**$p < .01$

answer this question also declined to answer most of the other questions.) The validation sample contained 64% supporters of the nuclear plant and 36% opponents, as compared with 69% and 31% in the primary sample. The difference of 5% is within the $1 - \alpha = .90$ sampling error calculated for the primary sample (see Sundstrom, Costomiris, DeVault, Dowell, Lounsbury, Mattingly, Passino, & Peele, 1977). The responses of the validation sample conform with our conclusion that about two-thirds of the residents of the county find the nuclear plant acceptable.

Expectancies and Subjective Values. The validation sample agreed very closely with the primary sample in ratings of expectancies and subjective values. For expectancy ratings, means in the validation sample were all within .5 standard deviation (and .5 scale step) of the means in the primary sample. The two groups of means were very highly correlated ($r = .93$; $n = 27$; $p < .001$). As with ratings of desirability, all 27 means in the validation sample were closer to 0 than the corresponding means in the primary sample. This result suggests that participants who had been interviewed before made somewhat greater use of the

TABLE 7.7
Multiple Regression Analysis of Acceptance Scores in the Primary Sample Using Only Likelihood Scores for Five Groups of Potential Effects as Predictors ($N = 261$)

Predictor	*Simple r*	*Multiple r*	r^2	*Change in* r^2	*F Value* ($df = 1,257$)
Hazards	.63**	.63	.40	.40	56.46**
Economic benefits	-.43**	.68	.47	.07	47.14**
Social disruption	.50**	.71	.51	.03	21.36**
Increased business[a]	-.23**	(.71)	(.51)	(0)	–
Outside attention[a]	-.24**	(.71)	(.51)	(0)	–

[a]Not in the regression equation.
**$p < .01$.

rating scales. However, average ratings in the validation sample were, for practical purposes, identical with those in the primary sample.

Traditional Expectancy-Value Model. For the validation sample, we calculated composite acceptance scores exactly as in the primary sample. (We used the means and standard deviations from the primary sample, which were practically the same as in the validation sample.) We also calculated Σ *EV* scores as in the primary sample. The correlation between acceptance scores and Σ *EV* scores was $-.64$ ($n = 80$; $p < .001$), reflecting a relationship that accounts for 40% of the variance in acceptance scores. The correlation in the validation sample was only slightly smaller than the correlation in the primary sample ($-.70$), and it shows strong support for the traditional approach to the expectancy-value model.

Factor-Score Approaches. Using the same groups of effects as in the primary sample, we repeated both factor-score approaches with the validation sample using multiple-linear-regression analysis. As shown in Table 7.8, the approach based on average Σ *EV* scores for the five groups of effects accounted for 46% of the variance in acceptance scores. Results were similar to those for the primary sample except that the second predictor was economic benefits instead of increased business. In effect, these two predictors switched places. However, the order of the economic predictors in the equation is relatively unimportant because of their strong correlations with each other and with acceptance scores.

The second factor-score approach employed only the mean expectancy scores for the five groups of effects. Results appear in Table 7.9. This approach did considerably better than either of the approaches based on Σ *EV* scores. It accounted for 52% of the variance in acceptance scores, using only two of the five predictors. The results show the predominance of two groups of effects: hazards and economic benefits.

TABLE 7.8
Multiple Regression Analysis of Acceptance Scores in the Validation Sample Using Likelihood x Desirability Scores for Five Groups of Effects as Predictors ($N = 80$)

Predictor	*Simple r*	*Multiple r*	*Multiple r^2*	*Change in r^2*	*F Value (df = 1, 76)*
Economic benefits	-.53**	.53	.28	.28	7.68*
Hazards	-.52**	.65	.42	.14	20.95**
Increased business	-.40**	.67	.46	.04	5.01*
Social disruption[a]	-.27*	(.69)	(.47)	(.01)	–
Outside attention[a]	-.41**	(.69)	(.47)	(0)	–

[a]Not in the regression equation.
*$p < .05$.
**$p < .01$.

TABLE 7.9
Multiple Regression Analysis of Acceptance Scores in the Validation Sample Using Only Likelihood Scores for Five Groups of Effects as Predictors (N = 78)

Predictor	*Simple* r	*Multiple* r	r^2	*Change in* r^2	*F Value* ($df = 1, 75$)
Hazards	.58**	.58	.34	.34	39.34
Economic benefit	-.52**	.72	.52	.15	28.53**
Outside attention[a]	-.27**	(.73)	(.54)	(.02)	–
Social disruption[a]	.24**	(.74)	(.54)	(0)	–
Increased business[a]	-.22*	(.74)	(.54)	(0)	–

[a]Not in the equation.
*$p < .05$.
**$p < .01$.

DISCUSSION

The study yielded two main findings. The first concerns residents' acceptance of the nuclear plant. As predicted, acceptance of the facility was strongly related to opinions about its consequences. Acceptance was greater to the extent that residents believed desirable effects likely and undesirable effects, unlikely. The most important potential effects of the nuclear plant were hazards and economic benefits. This finding emerged in both samples and was consistent for all three approaches. Results therefore strongly supported the expectancy value model. The refinement of the model incorporating groups of potential effects based on a factor analysis accounted for as much variance as the traditional approach, even using only E scores. The findings carry implications for the expectancy-value model and for the study of the social impacts of large-scale changes in a community.

Implications for the Expectancy-Value Model

Our earlier study (Sundstrom, Lounsbury, Schuller, Fowler, & Mattingly, 1977) suggested that those who supported the nuclear plant thought hazards unlikely but expected both the economic benefits of growth, such as more jobs, and the disruptive consequences of growth, such as overcrowding. We interpreted the responses of the supporters of the facility as a sign that they would make a "trade-off," in which they assumed that the nuclear plant was relatively safe and would tolerate undesirable effects in return for desirable ones. Those opposed to the nuclear plant expressed different expectations. They thought that hazards were more likely, social disruption more likely, and economic benefits less likely. In other words, they were less willing to assume that the facility was safe and were more unwilling to make a trade-off they saw in less favorable terms.

The present study showed similar results. The findings supported the idea that attitude toward the nuclear plant represents a balance of anticipated costs and benefits. Further, the results pointed to the same primary elements in the balance—hazards and economic benefits. Those who would accept the nuclear plant thought it would be relatively safe, and bring more economic benefits, with relatively little social disruption. Those who rejected it expected it to be less safe, less likely to bring benefits, and more likely to bring social disruption. The balance of costs and benefits seems to depend most on extremely desirable or undesirable effects.

The profile could have been different. The balance could have depended equally on all of the group of effects. Or the residents could have agreed about the extremely desirable or undesirable effects and made more varied judgments about the neutral effects. This would have made the neutral effects most important. But the most desirable and undesirable effects were apparently most critical.

This finding bears on a major assumption underlying the expectancy-value model advanced by Vroom (1964). The model explicitly treats all possible effects of an object as if they had equal importance. (The model gives equal weight to each of the expectancy-value combinations that form the $\Sigma\ EV$ scores.) Our results seem to contradict the assumption. Some types of effects were apparently more important than others. For example, hazards were quite important; outside attention was unimportant.

The identification of certain groups of effects as more important was based on the factor-score approach. Using fewer than 10 potential effects of the nuclear plant (in the groups labeled "hazards" and "economic benefits"), the factor-score approach accounted for as much variance in acceptance scores as did the traditional approach, which employed 27 effects. By implication, the factor-score approach was more efficient. This reinforces Fishbein and Ajzen's suggestion that "a small number of beliefs serve as the determinants of a person's attitude" (1975, p. 218); in fact, they estimate that in most cases, this number will be between five and nine salient beliefs. In future applications of expectancy-value theory, the identification of the most salient outcomes in a pilot study could result in a substantial reduction in the amount of information needed from participants in the research. The unwieldy lists of effects usually used in such research may be shortened, and ratings of the likelihood of effects may suffice without ratings of desirability.

Other implications also follow from the fact that ratings of likelihood accounted for as much variance in acceptance scores as the multiplicative scores based on Vroom's (1964) model. Our finding agrees with earlier research reviewed by Mitchell (1974). One possible explanation derives from the difficulties inherent in multiplicative scores based on ratings with unknown scale properties. For example, Schmidt (1973) has shown that the use of score transformations appropriate for interval-level data produces wide fluctuations in results related to expectancy theory. Perhaps the error introduced by the multiplicative scores countervails the information added by the ratings of desirability.

Another explanation for the results of the multiplicative scores derives from the possible lack of variability in one of the two component scores. For example, if the desirability ratings for an outcome tend not to vary, the multiplicative scores could represent little more than multiplication by a constant. This explanation can be ruled out. Ratings of both likelihood and desirability ranged over all possible values, and the standard deviations were substantial—usually about 1 scale step—and consistent. If anything, the multiplicative scores had greater possibilities for variation than the ratings of likelihood. However, the variation in the multiplicative scores could also reflect considerable error, because the errors of the two components are multiplied.

A third explanation for the relative effectiveness of the likelihood scores alone is that the ratings of subjective value added less unique information than the expectancy-value model assumes. The model implicitly assumes that the perceived likelihood of an effect is independent of its desirability. But the results of the present study contradict the assumption of independence. As ratings of desirability became more extreme, ratings of likelihood became somewhat higher.[4] The slight redundancy of the two types of ratings may have made the multiplicative scores less useful than they otherwise would be, independent of the amount of error they contain.

There was an exception to the general failure of the multiplicative scores to yield stronger relationships with acceptance scores—namely, the group of effects labeled "increased business." Effects in this group were rated "likely" and "desirable," as were the effects in the group we labeled "outside attention." But the correlations of the multiplicative scores on increased business and acceptance scores was about .40, compared with about .20 for the likelihood scores alone. Here the ratings of valence apparently did add enough information to overcome any potential error introduced by multiplication. Perhaps there are other cases in which multiplicative scores yield stronger predictions than likelihood ratings alone. However, in the present study, such cases clearly represented exceptions.

Implications for the Social Impacts of Large-Scale Projects

Perhaps the most important implication of the findings derives from the relatively large amount of variance in acceptance scores—about 50% accountable through the expectancy-value model. Despite fairly wide variations in levels of expectation and value concerning its specific effects, attitudes about the nuclear power plant seem to reflect a common cognitive process. The combination of elements

[4]It is interesting to speculate as to why expectancies and values are correlated. One interpretation might be that such a state of affairs represents attempts at cognitive balancing. For example, persons favoring the plant may exaggerate the desirability and likelihood of most valued outcomes while de-emphasizing the value and likelihood of negative outcomes.

in the manner of an expectancy-value model suggests that *rational processes* underlie decisions about acceptance of a nuclear power plant. This implication is at variance with the assertion that attitudes related to nuclear power are dominated by irrational processes (see Kasperson, Berk, Pijawka, Sharaf, & Wood, 1976).

Another implication of the findings for social-impact assessment of nuclear plants and other large-scale projects is that the projected impacts include the early formation of individual attitudes. The changes involved in such projects may produce subjective effects even before objective effects become evident. If subjective effects can be linked empirically to subsequent objective impacts, it may be possible to use data from studies like ours to predict objective social impacts. For example, perhaps disproportionate numbers of opponents of the nuclear plant will tend to leave the area soon after it is constructed or to vote against the construction of the plant in a public referendum. Such data might form the basis for a predictive model of social impacts that could prove useful to policy analysts and decision-makers. A predictive model could also figure prominently in programs for mitigation of adverse social impacts, such as the one introduced by the TVA in Hartsville (see Sundstrom, Costomiris, DeVault, Dowell, Lounsbury, Mattingly, Passino, & Peele, 1977). Thus, efforts could be made by the responsible agencies to alleviate in advance the potential impact of important "negative" outcomes predicted by the model (e.g., overcrowding in the schools or housing shortages).

In summary, the findings clearly supported the expectancy-value model and extended it through an approach that specified the elements of the cognitive "trade-off" that underlies individual attitudes toward a complex object. In this case, acceptance of a nuclear plant was associated with the expectation of economic benefits but also with social disruption. Results suggested that research on the expectancy-value model in the context of attitudes toward nuclear power may effectively employ shorter lists of outcomes and rely solely on ratings of likelihood. The results also indicated a strong rational decision-making process underlying attitudes toward a nuclear power plant. Data from studies like this one on subjective responses to large-scale projects may contribute to a predictive model of social impacts.

ACKNOWLEDGMENTS

This research was sponsored by the U.S. Energy Research and Development Administration, Division of Biomedical and Environmental Research, under Union Carbide Corporation's contract No. W-7405-eng-26 with the U.S. Energy Research and Development Administration. The research represents part of a program of investigations by the Social Impacts Analysis Group, Regional and Urban Studies Section, Energy Division, Oak Ridge National Laboratory.

Send requests for reprints to Eric Sundstrom, Department of Psychology, University of Tennessee, Knoxville, Tenn. 37916.

REFERENCES

Alwin, D. F. The use of factor analysis in the construction of linear composites in social research. *Sociological Methods & Research,* 1973, *2,* 191-214.

Finsterbusch, K., & Wolf, C. P. (Eds.). *Methodology of social impact assessment.* Stroudsburg, Pa.: Dowden, Hutchinson & Ross, 1977.

Fishbein, M., & Ajzen, I. *Belief, attitude, intention and behavior: An introduction to theory and research.* Reading, Mass.: Addison-Wesley, 1975.

Gorsuch, R. L. *Factor analysis.* New York: Saunders, 1974.

Kasperson, R., Berk, G., Pijawka, D., Sharaf, A., & Wood, I. *Nuclear energy, local conflict, and public opinion* (Review draft of technical report for Project R.A.R.E.—Risk Assessment of Rare Events). Worchester, Mass.: Clark University, August 1976. (Cited with permission.)

Lounsbury, J. W., Sundstrom, E., Schuller, C. R., Mattingly, T. J., Jr., & DeVault, R. C. Toward an assessment of potential social impacts of a nuclear power plant on a community: Survey of residents' views. In K. Finsterbusch & C. P. Wolf (Eds.), *The methodology of social impact assessment.* Stroudsburg, Pa.: Dowden, Hutchinson & Ross, 1977.

Mitchell, T. R., Expectancy models of job satisfaction, occupational preference and effort: A theoretical, methodological, and empirical appraisal. *Psychological Bulletin,* 1974, *81,* 1053-1077.

Mitchell, T. R., & Biglan, A. Instrumentality theories: Current uses in psychology. *Psychological Bulletin,* 1971, *76,* 432-454.

Motowidlo, S. J., Loehr, V., & Dunnette, M. D. The effect of goal specificity on the relationship between expectancy and task performance. *Journal of Applied Psychology,* 1972, *57,* 61-67.

Schmidt, F. L. Implications of a measurement problem for expectancy theory research. *Organizational Behavior and Human Performance,* 1973, *10,* 243-251.

Schuller, C. R., Fowler, J. R., Mattingly, T. J., Jr., Sundstrom, E. D., Lounsbury, J. W., Passino, E. M., Dowell, D. A., & Hutton, B. J. *Citizens' views about the proposed Hartsville nuclear power plant: Preliminary report of potential social impacts* (Report # ORNL-RUS-3). Oak Ridge, Tenn.: Regional and Urban Studies Section, Energy Division, Oak Ridge National Laboratory, May 1975.

Schuster, J. R., Clark, B., & Rogers, M. Testing portions of the Porter and Lawler model regarding the motivational role of pay. *Journal of Applied Psychology,* 1971, *55,* 187-195.

Sundstrom, E. D., Costomiris, L. J., DeVault, R. C., Dowell, D. A., Lounsbury, J. W., Mattingly, T. J., Jr., Passino, E. M., & Peele, E. *Citizens' views about the proposed Hartsville nuclear power plant: A survey of residents' perception in August, 1975* (Report # ORNL-TM-5801). Oak Ridge, Tenn.: Oak Ridge National Laboratory, 1977.

Sundstrom, E., Lounsbury, J. W., Schuller, C. R., Fowler, J. R., & Mattingly, T. J., Jr. Community attitudes toward a proposed nuclear power generating facility as a function of expected outcomes. *Journal of Community Psychology,* 1977, *5,* 199-208.

Susmilch, C. E., & Johnson, W. T. Factor scores for constructing linear composites (Do different techniques make a difference?) *Sociological Methods & Research,* 1975, *4,* 166-188.

Vroom, V. H. *Work and motivation.* New York: Wiley, 1964.

Wahba, M., & House, R. Expectancy theory in work and motivation: Some logical and methodological issues. *Human Relations,* 1974, *27,* 121-147.

Author Index

Italics denote pages with bibliographic information.

Subject Index